SHEPHERDS OF BRITAIN

SCENES FROM SHEPHERD LIFE
PAST AND PRESENT

FROM THE BEST AUTHORITIES

BY

ADELAIDE L. J. GOSSET

AULD KEP

Copyright © 2013 Read Books Ltd.
This book is copyright and may not be
reproduced or copied in any way without
the express permission of the publisher in writing

British Library Cataloguing-in-Publication Data
A catalogue record for this book is available from the
British Library

Sheep Farming

Sheep (*Ovis aries*) are quadrupedal, ruminant mammals, typically kept as livestock. Like all ruminants, sheep are members of the order 'Artiodactyla', the even-toed ungulates. Although the name 'sheep' applies to many species in the genus *Ovis*, in everyday usage it almost always refers to *Ovis aries*. Numbering a little over one billion, domestic sheep are also the most numerous species of sheep. A male sheep is called a 'ram' and a female sheep is called a 'ewe'.

There are four general styles of sheep husbandry to serve the varied aspects of the sheep industry - and the needs of a particular shepherd. Commercial sheep operations supplying meat and wool are usually either 'range band flocks' or 'farm flocks.' Range band flocks are those with large numbers of sheep (often 1,000 to 1,500 ewes) cared for by a few full-time shepherds, sometimes requiring the shepherds to live with the sheep as they move through the pasture. The pasture, which must be large enough to accommodate the sheep, can be either fenced or open and sheepdogs as well as means of transport such as horses or motor vehicles are commonly required. As range band flocks move within a large area in which it would be difficult to supply a steady source of grain, and consequently almost all subsist on pasture alone. This style of sheep raising accounts for most of the sheep operations in the U.S., South America and Australia.

Farm flocks are slightly smaller than range bands, and are kept on a more confined, fenced pasture land. Farm flocks may also be a secondary population on a larger farm, used by farmers who raise a surplus of crops to finish market lambs on or those with untillable land they wish to exploit. However, farm flocks account for many farms focused on sheep as primary income in the U.K. and New Zealand (due to the more limited land available in comparison to other sheep-producing nations). The farm flock is a common style of flock management for those who wish to supplement grain feed for meat animals. An important corollary form of flock management to the aforementioned styles are 'specialized flocks', raising purebred sheep. Many commercial flocks, especially those producing sheep meat, utilize cross-bred animals and breeders raising purebred flocks provide stud stock for these operations, often simultaneously working to improve the breed and participate in showing.

The last type of sheep keeping is that of the 'hobbyist'. This type of flock is usually very small compared to commercial operations, and may be considered pets. Those hobby flocks, which are raised with production in mind, may be for subsistence purposes or to provide a very specialized product, such as wool for handspinners. Quite a few people, especially those who have emigrated to rural areas from urban or suburban enclaves, begin with hobby flocks or a 4-H lamb before eventually expanding to farm or range flocks.

Sheep breeds are often grouped based on how well they are suited to producing a certain type of breeding stock. Generally, sheep are thought to be either 'ewe breeds' or 'ram breeds.' Ewe breeds are those that are hardy, and have good reproductive and mothering capabilities – they are for replacing breeding ewes in standing flocks. Ram breeds are selected for rapid growth and carcass quality, and are mated with ewe breeds to produce meat lambs. Lowland and upland breeds are also crossed in this fashion, with the hardy hill ewes crossed with larger, fast-growing lowland rams to produce ewes called 'mules', which can then be crossed with meat-type rams to produce prime market lambs. Many breeds, especially rare or primitive ones, fall into no clear category. Sheep breeders look for such traits in their flocks as high wool quality, consistent muscle development, quick conception rate (for females), multiple births, and quick physical development.

Keeping sheep can often be a complex business, as sheep have many natural predators – coyotes in North America, Foxes in Europe and dingoes in Australia. Lambs in pasture are particularly vulnerable, frequently falling prey to crows, eagles and ravens. Consequently, many sheep are kept in barns, paddocks and pens, as well as merely out in the open. Freshly shorn hoggets (a young sheep of either sex from about 9 to 18 months of age) can be very susceptible to wet and windy weather however, and are frequently kept inside. This can become hard to organise for the crofter, and consequently most Australian farmers arrange for all the ewes in a mob to give birth (the *lambing season*) within a period of a few weeks. As ewes sometimes fail to bond with new-born lambs, especially after delivering twins or triplets, it is important to minimize disturbances during this period. In order to more closely manage the births, vaccinate lambs, and protect them from predators, shepherds will often have the ewes give birth in 'lambing sheds'; essentially a barn (sometimes a temporary structure erected in the pasture) with individual pens for each ewe and her offspring. In Norway most of the ewes are examined with ultrasound equipment to determine how many lambs will be born.

Sheep are an important part of the global agricultural economy. However, their once vital status has been largely replaced by other livestock species, especially the pig, chicken, and cow. China, Australia, India and Iran have the largest modern flocks, and serve both local and exportation needs for wool and mutton. Other countries such as New Zealand have smaller flocks but retain a large international economic impact due to their export of sheep products. Despite the falling demand and price for sheep products in many markets, sheep have distinct economic advantages when compared with other livestock. They do not require the expensive housing, such as that used in the intensive farming of chickens or pigs. They are an efficient use of land; roughly six sheep can be kept on the amount that would suffice for a single cow or horse, and sheep can also consume plants such as noxious weeds that most other animals will not touch, and produce more young at a faster rate. Also, in contrast to most livestock species, the cost of raising sheep is not necessarily tied to the price of feed crops such as grain, soybeans and corn. Combined with the lower cost of quality sheep, all these factors combine to equal a lower overhead for sheep producers, thus entailing a higher profitability potential for the small farmer.

"To keep sheep the best life."—*Manningham Diary*, 1592-93.

"A shepherd's life, properly understood, is the richest in the world."—JAMES GARDNER, 1840-1900.

"The shepherd's life has changed less with the change of years than that of any other calling."—H. SOMERSET BULLOCK, 1909.

PREFACE

It is now many hundred years since sheep-farming became one of the first industries of Britain, and it is therefore surprising that no book at all adequately descriptive of the shepherds of this country and their shepherding has yet been published. The reason for this is perhaps the great scope of the subject, and the innumerable points of interest upon which such a book must necessarily touch, if only to set down and note the changes which even in this most conservative calling—the most unchanging of all our industries—have to be recorded. That the ways of shepherds and the charms of their life do interest us, I have ample evidence. For since the announcement was made of the preparation of this volume, I have received more than two hundred letters on the subject from correspondents in various parts of the country.

The scope of the book will appear from a glance at the contents list. It comprises chapters on "Shepherds—their Flocks and Dogs," "Sheep Marks and Tallies," "The Wool Harvest," "The Care of Wool," "Shepherds' Garb," "Arts and Crafts," "Pastimes," and "Pastoral Folk-Lore." Owing to the overwhelming abundance and richness of the material, I have made no attempt to do more than notice what are comparatively a few out of the many different breeds of sheep, though I have touched here and there upon some that for various reasons seemed especially noteworthy. Nevertheless the whole material has been arranged roughly according to counties, beginning with the south, where the conditions have been most generally modernised, and working gradually northward to Orkney and Shetland, where the shepherd has in many respects preserved his original customs, some of which, that of "rueing," for instance, probably date back to prehistoric times. In connexion with this it has been suggested that the black-horned sheep, with regard to which the reader will find many interesting particulars in the text, was probably the most important variety of sheep in prehistoric Britain. But what I chiefly felt in the work for this book was, after all, the deep-lying humanity of its subject, which reveals in the shepherd of real life, as clearly as in the shepherd of legend and history, the essentially regal qualities of insight and of sympathy.

To the writers of the letters mentioned above, a few my friends, but so many unknown to me, I desire to return hearty thanks for their kindly help. I must thank, too, those authors, artists, and publishers who have so generously given me permission to draw upon their work. Among many who have rendered important and substantial help, for which I wish to express my gratitude, are Mr. C. J. B. Macdonald of *The Field*; The Ven. Archdeacon Thomas, F.S.A.; Mr. J. C. Bacon; Miss A. Dryden; Miss Sophia Morrison; The Rev. Thomas Mathewson; Mr. Arthur Finn; Mr. W. B. Gardner; The Rev. F. W. Galpin; Mr. Edward Lovett; Mrs. Henry-Anderson; Mr. Ruskin Butterfield; Miss Charlotte Burne; Mr. Walter Money, F.S.A.; Dr. Chalmers Mitchell, F.R.S.; The Editor of *The Scottish Farmer*; Mr. F. C. Paine; Mr. D. Macpherson; Dr. S. Tellet; Mr. T. J. George; Mr. J. R. Farewell; Mr. Merrick Head; Mr. A. Beckett; Prof. Skeat; and lastly, to Mr. Walter Skeat, M.A., for several articles, and for much other kind and valuable assistance.

A. L. J. GOSSET.

CONTENTS

SHEPHERD AND FLOCK

SUSSEX AND HAMPSHIRE—

SHEPHERDS OF THE DOWNS. By R. W. BLENCOWE . 3
    From *Sussex Archaeological Collections*, vol. ii., 1849.

NATURE AND THE COUNTRYMAN. By TICKNER EDWARDES* 5
    From *The West Sussex Gazette*, 1907.*

CONTENTEDNESS OF SOUTHDOWN SHEPHERDS. By W. H. HUDSON* 11
    From *Nature in Downland*, 1900. (Messrs. Longmans, Green & Co.)*

VASTNESS OF SOUTHDOWN SHEEP-WALKS.
                    By RICHARD JEFFERIES 13

A DOWNLAND SHEEP FAIR.   ,,   ,,   14
    From *Nature near London*, 1887. (Messrs. Chatto & Windus.)*

SHEPHERDS' HUTS ON THE SOUTH DOWNS. By M. A. LOWER 17
    From *Contributions to Literature*, 1854.

A SHEPHERD'S BUSH. By WODEHOUSE R. H. GARLAND* 18
    From *Morning Post*, 1910.*

JOHN DUDENEY.   By W. H. HUDSON* . 19
A SUSSEX SHEPHERDESS.   ,,   ,,   20
A PICTURESQUE SHEPHERD-BOY.   ,,   ,,   23
    From *Nature in Downland*, 1900. (Messrs. Longmans, Green & Co.)*

FROM SHEPHERD-LAD TO LAND-HOLDER. By J. BATEMAN* 24
    From *The West Sussex Gazette*, 1907.*

            * By kind permission.

Shepherds of Britain

	PAGE
A HAMPSHIRE SHEPHERD. By H. RIDER HAGGARD*	26
From *Rural England*, 1902. (Messrs. Longmans, Green & Co.)*	

DORSETSHIRE—
HARDINESS OF THE PORTLAND BREED. By THE AUTHOR 27

DEVONSHIRE—
COURAGE OF THE EXMOOR SHEEP. By J. H. CRABTREE* 28
From *The Animal World*, 1909.*

CORNWALL—
SHEEP OF THE "TOWENS" AND SCILLY. By WILLIAM BORLASE, F.R.S. 28
From *The Natural History of Cornwall*, 1758.

SOMERSETSHIRE—
BEN BOND, IDLETON. By JAMES JENNINGS, 1834 . 29

WILTSHIRE—
THE SHEPHERD OF THE PLAIN. By PERCY W. D. IZZARD* 34
From *The Daily Mail*, 1910.*

LAZY SHEPHERDS OF THE PLAIN—AND AN EXCEPTION. By THE AUTHOR 36

WILTSHIRE SHEPHERDS, 17th Century . . . 37
From *The Book of Days*. Edited by ROBERT CHAMBERS, 1869.

WILTSHIRE SHEPHERD CUSTOMS. By RICHARD JEFFERIES 40
From *Wild Life in a Southern County*, 1880. (Messrs. Smith, Elder & Co.)*

CHARACTERISTIC WILTSHIRE SHEPHERDS. By A. G. BRADLEY 43
From *Round about Wiltshire*, 1907. (Messrs. Methuen & Co.)*

BERKSHIRE—
SHEPHERDING ON THE BERKSHIRE DOWNS. By J. E. VINCENT 46
From *Highways and Byways in Berkshire*, 1906. (Messrs. Macmillan & Co.)*

* By kind permission.

Contents

	PAGE
A SHEEP FAIR AT EAST ILSLEY. By L. SALMON*	46
From *Untravelled Berkshire*, 1909. (Messrs. Sampson Low, Marston & Co.)*	

KENT—

A SHEPHERD'S POWER OF ABSTRACTION. By RICHARD JEFFERIES	49
From *Nature near London*, 1887. (Messrs. Chatto & Windus.)*	
"SHEPPEY," THE ISLE OF SHEEP. By WILLIAM LAMBARDE	50
From *Perambulations in Kent*, 1576.	
THE SHEEP OF ROMNEY MARSH	51
From *Excursions in the County of Kent*, 1822.	
THE "LOOKERS" OF ROMNEY MARSH. By ARTHUR FINN, 1910	51

EAST ANGLIA—

THE NORFOLK BREED. By THE AUTHOR	53

LINCOLNSHIRE—

LINCOLNSHIRE LONGWOOLS. By THE AUTHOR	55

THE MIDLANDS—

"LEICESTERS" AND OTHERS	55
From *Pictorial Half-Hours*. Edited by CHARLES KNIGHT, 1850.	

YORKSHIRE—

HENRY CLIFFORD, "THE SHEPHERD LORD"	56
From *The Book of Days*. Edited by ROBERT CHAMBERS, 1869.*	
SHEPHERDS' MEETING-TIME	59
From *The Globe*, 1878.*	

THE LAKE COUNTRY—

SHEEP-FARMING IN CUMBERLAND. By J. BRITTON and E. W. BRAYLEY	60
From *The Beauties of England and Wales*, 1802.	
THE HERDWICK. By A. G. BRADLEY	60
A CURIOUS USAGE IN THE LAKE DISTRICT. By A. G. BRADLEY	61
From *Highways and Byways in the Lake District*, 1901. (Messrs. Macmillan & Co.)*	

* By kind permission.

	PAGE
ANOTHER ACCOUNT OF THE HERDWICK	62
From *The Morning Post*, 1909.*	
ISLE OF MAN—	
THE TWO BREEDS OF THE ISLAND SHEEP. By JOHN FELTHAM	62
From *A Tour in the Isle of Man*, 1798.	
THE "LAUGHTONS" OR "LOAGHTANS"	63
From *The Sheep*, 1837.	
SHEEP AND SHEPHERDING IN THE ISLE OF MAN. By THE AUTHOR	63
OLD ENGLISH BREEDS OF SHEEP. By WALTER SKEAT, M.A., 1910	67
WALES—	
SHEPHERDING IN ANCIENT WALES	69
From *Ancient Welsh Husbandry* (A Commercial and Agricultural Magazine).	
HABITS OF WELSH SHEEP. By THEOPHILUS JONES	69
From *A History of the County of Brecknock*, 1805-1809.	
SHEEP CHARACTER IN CARNARVONSHIRE. By the Rev. J. EVANS	71
From *The Beauties of England and Wales*, 1812.	
SHEEP OF THE SNOWDONIAN RANGE. By THE AUTHOR	72
IRELAND—	
SHEPHERD AND FLOCK IN ERIN. By RALPH FLEESH, 1909	73
SHEEP IN ANCIENT IRELAND. By THE AUTHOR	76
CONNEXION BETWEEN THE IRISH AND FAROES' BREED. By THE AUTHOR	77
SCOTLAND—	
MY HIGHLAND SHEPHERD FRIENDS. By KATE HENRY-ANDERSON, 1909	79
HARDSHIPS OF SHEPHERD LIFE IN THE HIGHLANDS. By ALEXANDER INNES SHAND	81
From *Days of the Past*, 1905. (Messrs. Constable & Co., Ltd.)*	
JAMES GARDNER, SHEPHERD AND FAMOUS COLLIE-DOG TRAINER	84
An Appreciation by a Grateful Pupil, 1909.	

* By kind permission.

Contents

	PAGE
JAMES GARDNER—ANOTHER ACCOUNT. By the Rev. HUGH YOUNG, 1910	85
SAYINGS ON DOGS, TAKEN FROM THE CONVERSATION OF JAMES GARDNER	88
ON SHEPHERDS AND SHEPHERDING IN SKYE. By ALEXANDER SMITH	89
From *A Summer in Skye*, 1865.	
THE BLACKFACE BREED. By D. MACPHERSON, 1909	91
DEER EXPELLED BY SHEEP	92
From Hone's *Table Book*, 1827.	
SCOTTISH SHEPHERDS OF THE SIXTEENTH CENTURY. By JAMES TAYLOR, D.D.	93
From *The Pictorial History of Scotland*, 1859.	
THE MILKING OF EWES. By THE AUTHOR	95

SHETLAND AND ORKNEY ISLES—

THE WILD SHEEP OF SHETLAND. By Dr. S. HIBBERT	97
From *A Description of the Shetland Isles*, 1822.	
SHETLAND SHEEP—ANOTHER ACCOUNT. By ROBERT COWIE, M.D.	98
From *Shetland, Descriptive and Historical*, 1871. (Messrs. Lewis Smith & Son.)*	
SHEPHERDING IN SHETLAND AND ORKNEY. By the Rev. T. MATHEWSON, 1909	99
BREEDS OF SHEEP IN SHETLAND AND ORKNEY. By JAMES JOHNSTON, 1909	101

RARER PHASES OF SHEEP CULTURE AND CHARACTER

SHEEP LED BY THE SHEPHERD. By THE AUTHOR	105
SHEPHERDESSES OF THE SEVENTEENTH CENTURY. By THE AUTHOR	106
AN OLD-TIME LINCOLNSHIRE DROVER	107
From *Evening News*, 1908.*	
THE LITTLE NORTHAMPTONSHIRE DROVER	108
From *The Gentleman's Magazine*, 1797.	

* By kind permission.

	PAGE
THE INFLUENCE OF ENVIRONMENT. By THE AUTHOR	109
THE "FLEECY RACHAEL" WEEPING FOR HER CHILDREN. By ALEXANDER SMITH	111
From *A Summer in Skye*, 1865.	
MUTUAL RECOGNITION BY SHEEP AFTER SHEARING. By THE AUTHOR	112
ON PASTURE POISONOUS TO SHEEP. By THE AUTHOR	112
DEPENDANCE OF SHEEP ON THE WEATHER FOR FOOD AND DRINK. By RICHARD JEFFERIES	114
From *Wild Life in a Southern County*, 1880. (Messrs. Smith, Elder & Co.)*	
THE SNAIL-EATER. By H. L. F. GUERMONPREZ*	116
From *The West Sussex Gazette*, 1910.*	
THE BONE-EATER. By THE AUTHOR	118
THE BLIND SHEEP	119
From *Sunday*, 1907. (Messrs. Wells, Gardner, Darton & Co.)*	
A "MODERATE" DRINKER. By THE AUTHOR	119
A SHEEP MILITANT. By E. B. H.*	120
From *Country Life*, 1905.*	

THE SHEPHERD AND HIS DOG

SHEEP-DOGS, PAST AND PRESENT. By WALTER BAXENDALE, 1909	125
THE OLD ENGLISH OR SUSSEX SHEEP-DOG. By THE AUTHOR	127
THE MEANING OF "COLLIE." By THE AUTHOR	129
THE COLLIE-DOG OF THE HIGHLANDS. By K. HENRY-ANDERSON, 1909	130
THE COLLIE IN THE SOUTH OF SCOTLAND. By Sir ARCHIBALD GEIKIE, K.C.B., LL.D., D.C.L.	131
From *Scottish Reminiscences*, 1904. (Messrs. James MacLehose & Sons.)*	
THE POWERS OF THE COLLIE. By CHARLES ST. JOHN	133
From *Wild Sports in the Highlands*, 1846.	

* By kind permission.

Contents

	PAGE
HOW MASTER AND DOG CO-OPERATE. By RALPH FLEESH, 1909	135
HOGG'S "FAITHFUL SIRRAH AND HECTOR." By JAMES HOGG ("The Ettrick Shepherd," b. 1772, d. 1835). From *Anecdotes of Dogs* (Jesse), 1846.	140
THE SHEEP-DOG OF IRELAND. By RALPH FLEESH, 1909	142
SHEEP-DOGS (*coill*) IN THE ISLE OF MAN. By THE AUTHOR	143
THE SHEPHERD'S AND DROVER'S DOGS COMPARED. From *Pictorial Half-Hours*, 1851.	145
THE DROVER'S DOG. By EDWARD JESSE. From *Scenes and Occupations of a Country Life*, 1853.	145
TRAINING THE SHEEP-DOG IN ENGLAND. By H. SOMERSET BULLOCK, B.D., 1909*	147
TRAINING THE COLLIE PUP IN SCOTLAND. By RALPH FLEESH, 1909	148
SHEEP-DOG TRIALS IN ENGLAND, WALES, AND SCOTLAND. By WALTER BAXENDALE, 1910	149
A LIGHT OF OTHER DAYS. By C. BREWSTER MACPHERSON* From *The Field*, 1909.*	152
AULD KEP: "A PAST-MASTER," AND "ONE OF THE GREAT DOGS OF HISTORY." By RALPH FLEESH*. From *The Field*, 1909.*	154
A SCOTTISH SHEEP-DOG TRIAL. Reported in Doric. By RALPH FLEESH, 1909* From *People's Journal.*	155
"MAGNUS" AND "RONALD." By MAX PHILPOT, 1909 * (Messrs. W. M. Peace & Son, Orkney.)*	160
SHEPHERDS' DOGS IN CHURCH. By THE AUTHOR	167
SHEPHERDS' DOGS EXPELLED FROM CHURCH. By the Ven. ARCHDEACON THOMAS, F.S.A.* From *Archaeologia Cambrensis.*	170
OTHER METHODS OF EXPULSION. By THE AUTHOR	173

* By kind permission.

SHEPHERDS' DOGS AND SHEEP-STEALING. By JAMES
HOGG ("The Ettrick Shepherd") . . . 176
(As quoted by Jesse, 1846.)

SHEEP-MARKS AND TALLIES

ON SHEEP-MARKING. By THE AUTHOR . 181
LAMB-BRANDING IN SKYE. By ALEXANDER SMITH . 185
From *A Summer in Skye*, 1865.
SHEEP-MARKS. By the Rev. THOMAS MATHEWSON, 1909 188
EAR-MARKING IN SHETLAND. By Dr. JAKOBSEN, 1909* 190
TALLY-STICK REGISTERS. By EDWARD LOVETT* . 191
From *Folk-Lore*, 1909.*
NOTCHES AND NICKS. By THE AUTHOR . 193
THE SHEEP-COUNTING SCORE. By WALTER SKEAT,
M.A., 1910 194

WOOL HARVEST

THE WASHPOOL. By RICHARD JEFFERIES . . 203
From *Wild Life in a Southern County*, 1880. (Messrs. Smith, Elder & Co.)*
A SUSSEX SHEEP-SHEARING. By R. W. BLENCOWE . 206
From *Sussex Archaeological Collections*, 1849.
A SHEEP-SHEARING SONG TO THE TUNE OF "ROSEBUDS
IN JUNE." By THE AUTHOR 210
SHEEP-WASHING: AN OLD INSTITUTION NOW DECLINING 211
From *The Morning Post*, 1909.*
SHEEP WASHING AND SHEARING. By JOHN TIMBS . 212
From *Nooks and Corners of English Life*, 1867.
HOLKHAM: A FAMOUS SHEEP-SHEARING FEAST. By
A. M. STIRLING 214
From *Coke of Norfolk*, 1908. (Mr. John Lane.)*
THE SHEARERS' KING AND QUEEN. By WILLIAM
HOWITT 216
From *The Book of the Seasons*, 1831.

* By kind permission.

Contents xvii

A CUMBERLAND CLIPPING. By A. W. R. . . 218
   From *The Table Book of William Hone*, 1826.

ON "RUEING" *VERSUS* CLIPPING. By THE AUTHOR . 219

THE CARE OF WOOL
THE LABOURS OF THE LOOM

OF WOOL AND WOOLLEN CLOTH. By the Very Rev. DANIEL ROCK, D.D. 225
   From *Textile Fabrics Art Handbook*, 1876. (The Board of Education.)*

WOOLLEN MANUFACTURE IN ENGLAND, 15th Century . 228
   From *The Antiquary's Portfolio*, 1825.

THE STORY OF THE COTSWOLD WOOL TRADE. By FRANCIS DUCKWORTH, M.A.* . . . 229
   From *The Cotswolds*, 1908. (Messrs. Adam & Charles Black.)*

KENDAL "COTTONS" 232
   From *The Reliquary*, 1861.

"CORNISH HAIR." By WILLIAM BORLASE, F.R.S. . 234
   From *The Natural History of Cornwall*, 1758.

THE GREY CLOTHS OF KENT. By JOHN TIMBS. . 234
   From *Nooks and Corners of English Life*, 1867.

"KENDAL GREEN," "COVENTRY BLUE," "LEOMINSTER ORE," "LINCOLN GREEN," AND "BRISTOL RED." By THE AUTHOR 235

THE GREAT FAIR OF STOURBRIDGE. By JAMES E. THOROLD ROGERS 237
   From *History of Agriculture*, 1882. (The Delegates of the Clarendon Press.)*

"SHEPHERD'S PLAID." By J. R. PLANCHÉ . . 238
   From *The History of British Costume*, 1834.

SHETLAND WOOL. By ROBERT COWIE, M.D. . . 238
   From *Shetland, Descriptive and Historical*, 1871. (Messrs. Lewis Smith & Son.)*

* By kind permission.

xviii   Shepherds of Britain

PAGE

WOOLLEN CLOTHS OF IRELAND 241
From *The Antiquary's Portfolio*, 1825.
A FAMOUS COAT-MAKING WAGER. By WALTER MONEY, F.S.A. 241

SHEPHERDS' GARB

SHEPHERDS' DRESS, PAST AND PRESENT. By THE AUTHOR 245
SMOCKS AND THEIR WEARERS. By WILLIAM HOWITT 251
From *A Country Book*, 1831.

SHEPHERDS' ARTS, IMPLEMENTS, AND CRAFTS

THE PASTORAL CROOK. By RICHARD JEFFERIES . 255
From *Nature near London*, 1887. (Messrs. Chatto & Windus.)*
SHEPHERDS' CROOKS. By E. V. LUCAS . . . 257
From *Highways and Byways in Sussex*, 1904. (Messrs. Macmillan & Co.)*
HOOK AND CROOK. By THE AUTHOR . . 257
OF SHEPHERDS' SHEARS. By THE AUTHOR . 259
SHEPHERDS' SLINGS. By JOSEPH STRUTT . . 260
From *The Sports and Pastimes of the People of England*, 1801.
WHEATEAR TRAPPING BY SHEPHERDS. By ARTHUR BECKETT * 261
From *The Spirit of the Downs*, 1909. (Messrs. Methuen & Co.)*
OF "EARTH-STOPPING" BY SHEPHERDS. By H. SOMERSET BULLOCK, 1910 * 264
OF THE SHEPHERD'S BOTTLE. By THE AUTHOR . 264
SHEEP-BELLS, ANCIENT AND MODERN. By THE AUTHOR 265
THE SIMPLE SUNDIAL OF THE SOUTHDOWN SHEPHERDS. By EDWARD LOVETT * 272
From *Folk-Lore*, 1909.

* By kind permission.

Contents

	PAGE
AN ANCIENT RUSTIC POCKET-DIAL. By THOMAS QUILLER COUCH	274
From *The Reliquary*, 1862.	
A SHEPHERD'S POCKET-DIAL. By E. B.* .	276
From *Country Life*, 1904.*	

SHEPHERDS' PASTIMES

A PIPING LAD. By RICHARD JEFFERIES . .	282
From *Wild Life in a Southern County*, 1880. (Messrs. Smith, Elder & Co.)*	
SHEPHERDS' PIPES. By Rev. F. W. GALPIN, 1909 .	282
SHEPHERDS AT PLAY	285
From *The Graphic and Historical Illustrator*, 1834. Edited by E. W. BRADLEY.	
SHEPHERDS OF SKYE AND THE REEL OF HOOLICAN. By ALEXANDER SMITH	286
From *A Summer in Skye*, 1865.	
THE COTSWOLD GAMES	287
From *The Book of Days*. Edited by ROBERT CHAMBERS, 1869.*	
SHEEP-RUNNING ON EXMOOR. By PERCY W. D. IZZARD*	290
From *The Daily Mail*, 1910.*	
VILLAGE PASTIMES IN THE SEVENTEENTH CENTURY .	290
From *The Book of Days*. Edited by ROBERT CHAMBERS, 1869.*	
THE OLD BERKSHIRE REVELS. By L. SALMON* .	291
From *Untravelled Berkshire*, 1909. (Messrs. Sampson Low, Marston & Co.)*	
BEDFORDSHIRE SHEARING REVELS .	292
From Hone's *Year Book*, 1832.	
ST. BLAISE'S DAY IN YORKSHIRE . . .	294
From *The Book of Days*. Edited by ROBERT CHAMBERS, 1869.*	
OLD CUSTOMS AT SHEPHERDS' FESTIVALS .	296
From *The Graphic and Historical Illustrator*, 1834.	

* By kind permission.

xx   Shepherds of Britain

	PAGE
NINE MEN'S MORRIS AND OTHER GAMES. By THE AUTHOR	297

 Quoting from *Traditional Games*. By ALICE GOMME, 1894.*
  ,, ,, *Notes and Queries*, 1878.*
  ,, ,, *Barnes' Glossary*, 1864. (Messrs. Kegan Paul, Trench, Trubner & Co.)*
  ,, ,, *Sports and Pastimes*. By JOSEPH STRUTT, 1801.

THE GAME OF JACK STRAWS. By WILLIAM HOWITT 301
 From *The Boy's Country Book*, 1841.

PASTORAL FOLK-LORE

THE SHEPHERD AND HIS LORE. By Dr. HABBERTON LULHAM 305
 From *Songs of the Downs and Dunes*, 1908. (Messrs. Kegan Paul, Trench, Trubner & Co.)*

THE PHYNODDEREE'S SHEPHERDING. By SOPHIA MORRISON, 1909 308

THE LOAGHTAN BEG. By "CUSHAG" (J. KERMODA)*. 308

REMNANTS OF SACRIFICIAL CUSTOMS IN ENGLAND. By WILLIAM HENDERSON 309
 From *Notes on the Folk-Lore of the Northern Counties of England*, 1879. (Folk-Lore Society's Publications.)*

SACRIFICE OF SHEEP AND LAMBS. By the Rev. J. E. VAUX, F.S.A. 310
 From *Church Folk-Lore*, 1902. (Messrs. Skeffington & Son.)*

SACRIFICIAL CUSTOMS AND OTHER SUPERSTITIONS IN THE ISLE OF MAN. By SOPHIA MORRISON, 1910 . 311

CHARMS AND CURE OF DISEASE BY MEANS OF SHEEP. By THE AUTHOR 313
 From *Folk-Lore*, 1902, 1908.*
 From *Notes and Queries*.*

A SHEPHERD BURIAL CUSTOM. By THE AUTHOR . 314

WEATHER WISDOM OF THE SHEEP. By THE AUTHOR 315

INDEX 321

   * By kind permission.

ILLUSTRATIONS

	PAGE
Cover Illustration. By Dr. Habberton Lulham.	
Lambs. By Dr. Habberton Lulham. *Frontispiece in Photogravure*	
"Auld Kep." The property of Mr. James Scott of Ancrum (Roxburghshire).* From *The Field* * *Title Page*	
"When the Weald of Sussex was full of Iron Mines." From *Sussex Archaeological Collections,* vol. xlvi. S.A.S.*	2
Born to the Craft. Photograph by Dr. Habberton Lulham	9
Springtime in the Sussex Weald. Photograph by Dr. Habberton Lulham	15
"Southdowns" : "Models of what Hill Sheep should be." Photograph by Dr. Habberton Lulham . .	25
A Shepherd's Care. Photograph by Dr. Habberton Lulham	39
Mothering. Photograph by the Rev. A. H. Blake .	41
Backstays. Photograph by Edward Lovett* . .	52
Norfolk Ewes and Lambs. A Ram of the Flock. {The property of Mr. Dermot M'Calmont. Photograph by Clarence Hailey & Co., Newmarket.* Lent by Mr. F. C. Paine*}	54
Loaghtan Ewe and Ram. From a Painting by W. Marsden. The property of Mr. J. C. Bacon * .	65
Mr. J. C. Bacon and Loaghtan Sheep * . . .	67

* By kind permission.

	PAGE
St. Kilda Sheep. The property of Mr. D. Macpherson*	68
A Three-Horned Sheep. From *The History of Quadrupeds*, 1781	78
Mr. James Gardner, Shepherd and Famous Collie Trainer*	84
Called to the Troughs. Photograph by C. Reid, Wishaw, N.B. From *Country Life*.	91
On the South Downs above Fulking. Photograph by Dr. Habberton Lulham	110
A Rainstorm. From a Painting by R. Westall, R.A. Engraved by R. M. Meadows, 1813	115
"Mootie," a Shetland Collie. Bred by Mr. A. J. Jamieson of Scalloway, Shetland (Owner)*	125
"Bob," a Sussex Sheep-Dog. Photograph by Mr. Stewart Acton (Owner)*	128
"Frisk" (a short-haired "beardie") and Mr. Alexander Millar of Burnfoot, Ayrshire (Owner).* Photograph by A. Brown of Lanark*	139
"Ben" (a rough grey merle) and Mr. Thomas Gilholm (Owner)*	151
Mr. W. B. Gardner (Ralph Fleesh), Sheep-Dog Judge*	157
Dog Tongs. From Penmynydd Church, Anglesey. By permission of the Rev. Morris Griffith	171
Dog Tongs. From Llaneilian Church, Anglesey. By permission of the Rev. J. J. Ellis	171
Dog Tongs. From Bangor Cathedral	171
Dog Tongs. From Clodock Church, Herefordshire. By permission of the Ven. Archdeacon Thomas and the Editor of *Archaeologia Cambrensis*	172
Dog-Whip, Paten, and Wooden Collecting-Box in Baslow Church, Derbyshire. Photograph by Mr. J. C. Sands. Copyright, Mr. A. Coates, Baslow*	173
"Old Scarleit," the Dog-Whipper. From a life-size Painting in Peterborough Cathedral. From a Lithograph belonging to Mr. G. C. Caster of Peterborough*	175

* By kind permission.

Illustrations

	PAGE
The Badge of Ownership. Photograph by Dr. Habberton Lulham	181
Ear-Marks. By the Rev. Morris Griffith*	185
Shetland Ear-Marks. By Dr. Jakobsen*	190
A Lamb Tally. By Edward Lovett*	191
Tally-Stick Registers. By Edward Lovett.* From *Folk-Lore*.*	192-193
Sheep-Washing in Sussex. From a Painting by J. Aumonier, R.I.* W. A. Mansell & Co., Photograph	205
Shearing Time. From a Painting by H. Singleton. Engraved by Cardon, 1801	210
The Delicious Downs of Albion. Photograph by Dr. Habberton Lulham	225
"Shepherd's Plaid." From an Original Sketch, 1851	239
A Shepherd of 1836. Wearing Buskins, and Sheepskin Cloak strapped plaidwise across the Shoulder	245
Bob Pennicott in Short Smock, carrying "Bottle," Bell, and Crook. "Jack" in Attendance	249
Shepherd Smith of Washington, near Chichester, in Smock and "Chummey"	250
The Wood-Carver. From a Painting by Alfred Parsons, A.R.A.* W. A. Mansell & Co., Photograph	256
A Pyecombe Hook	257
Shears represented on a Tombstone. From *Half-Hours with English Antiquities*, 1880	259
A "Music-Maker." Photograph by Dr. Habberton Lulham	266
Figures of Sheep-Bells. By Sir Henry Dryden, Bart. From Drawings in a Letter preserved in the British Museum*	267
Some Sheep-Bells in Possession of the Author	270
Sheep-Bells, Ancient and Modern	271
The Fortingal "Saint's Hand-Bell"	271

* By kind permission.

	PAGE
Turf Sundials. Photographs by Edward Lovett.* From *Folk-Lore*	272-273
A Pocket-Dial. From *The Reliquary*, 1862 . .	275
Shepherd with Pipe and Dog. From *A Book of Hours*, 1410	281
A Lesson in Piping. By H. Warren . . .	283
Returning to the Fold. By H. W. B. Davis, R.A.* W. A. Mansell & Co., Photograph . . .	296
Merrelles Board, Fourteenth Century. From Strutt's *Sports and Pastimes* . . .	298
The Shepherd. Photograph by Dr. Habberton Lulham .	305
Fleeces of Sky and Land. Photograph by Dr. Habberton Lulham	317

* By kind permission.

SHEPHERD AND FLOCK

From "Sussex Archaeological Collections," vol. xlvi. *By permission S. A. S.*

"AT THE TIME WHEN THE WEALD OF SUSSEX WAS FULL OF MINES"
A Sussex iron fire-back, 15th or early 16th century, bearing figures of sheep.

"Very different in form and symmetry was the sheep of those days from the beautiful animal which is now the pride and boast of Sussex."

SUSSEX AND HAMPSHIRE

SHEPHERDS OF THE DOWNS

By R. W. BLENCOWE, 1849

AT the time when, in the words of Camden, "the Weald of Sussex was full of iron mines, and the beating of hammers upon the iron filled the neighbourhood round about with continual noise," another large portion of the county, that of the South Downs, was, perhaps, one of the most solitary, noiseless districts in England. Princely Brighton was only a village of fishermen; Worthing a hamlet of another village, that of Broadwater; and within its boundaries there was but one town, that of Lewes, which really belonged to it. Here and there only, as is testified by maps of comparatively very recent date, along its southern slopes, or in the bottom of its valleys, was the land under tillage; over all the rest were spread vast flocks of sheep, which, with their attendant shepherds, ranged over a thousand breezy hills.

Few people, probably, are aware of the immense number of sheep which, under the twofold impulse of foreign demand and that given to it by the great woollen manufacture at home, were reared in England at an early period of our history. A large exportation of English sheep to Spain took place as early as 1273, in the reign of Alonzo X., when they were first imported there. According to a modern Spanish writer, Copmany, they were again imported in 1394, in the reign of Henry III. of Spain, as a part of the marriage portion of his wife,

Catherine Plantagenet, daughter of John of Gaunt; and Holinshed tells us, that "on the occasion of a treaty of alliance between Edward IV. of England and Henry IV. of Castille, license was given for certain Cotteswolde sheep to be transported into the countrye of Spaine, which have there so multiplyed and increased, that it hath turned the commoditie of England much to the profite of Spayne." "Above all," says an Italian writer in the year 1500, "the English have an enormous number of sheep, which yield them wool of the finest quality"; and we learn from an old record in the Exchequer, that in the 28th year of Edward III., in 1354, there were exported 31,651 sacks of wool and 3036 cwt. of fells. "In 1551 no fewer than sixty ships sailed from the port of Southampton only, laden with wool for the Netherlands." But that which throws the strongest light upon this point is a statute of the 29th Henry VIII., showing to what an extent the pasturage of the flocks had superseded the tillage of the land. The following is an extract:—"One of the greatest occasions that moveth and provoketh greedy and covetous people so as to accumulate and keep in their own hands such great portions of the land of this realme from the occupying of poor husbandmen, and so to use it in pasture, and not in tillage, is only the great profit that cometh of sheep. . . . So that some have 24,000, some 20,000, some 10,000, some 5000, some more, some less, by the which a good sheep for victual, that was accustomed to be sold for 2s. and 4d., or 3s. at most, is now sold for 6s., 5s., or 4s., or for 3s., and a stone of clothing wool, that in some shires was accustomed to be sold for 18d. or 20d., is now sold for 4s. and 3s. and 3d. at the least"; and then it enacts that no tenant occupier shall keep more than 2000 sheep exclusive of lambs under a year old. This large conversion of pasture lands into tillage accounts for the ridges and furrows which we see so frequently in grass fields.

Very different in form and symmetry was the sheep of those days from the beautiful animal which is now

Shepherd and Flock

the pride and boast of Sussex. The flocks were then reared more for their fleeces than their flesh. The wool trade, which had greatly advanced under the encouragement given to it by Edward III., went on improving and extending itself under many succeeding reigns, until it became the great staple manufacture of England. In Henry the VII.'s time it had established itself for the coarser manufactures in Yorkshire, particularly at Wakefield, Leeds, and Halifax ; and in the reign of Elizabeth it was firmly fixed in the west of England, where all the finer manufactures were, and indeed still are carried on. Its influence on the social and political condition of the people was very great : wealth flowed in, towns and villages were created by it, prices rose, rents increased, labour became more valuable, and gradually the middle. and lower classes of the people took a higher place in the social scale. When John Winchcomb, the clothier, commonly known by the name of Jack of Newberry, sent forth a hundred men, armed and clothed at his own expense, to meet the Scots at Flodden Field, the feudal baronial system had been shaken to its centre, and the loom was one of the most powerful of the levers which overthrew it. Independently of higher associations, there is a peculiar interest attached to the shepherd and his flock, and indeed to his faithful dog, arising from the general solitude of his life, from the scenery, particularly on the South Downs, in which he moves, and from the importance of his charge ; and, under the influence of this feeling, it seemed desirable to collect and preserve any old customs and habits connected with his mode of life which have passed, or which are about to pass, away.

NATURE AND THE COUNTRYMAN

By TICKNER EDWARDES, 1907

If the philosophers are right in attributing to environment such an enormous influence in the making or marring of men, why, it has been often asked, should the average agricultural labourer be such a stolid, undiscerning person? He lives and works, generally speaking, in the midst of the most beautiful surroundings. The wonderful life of field and woodland lies at his very door. All day long he can feast eyes and ears, if he will, on things that most of us would travel a dozen miles to hear and see. And yet he seems to let all go by him and over him unheeded; his thoughts take no flight, apparently, beyond the narrow horizon of his daily work, or the care of his allotment garden; and he finds pleasure seemingly in little else than his meals and the evening hour at the inn.

But the truth is, under modern conditions at least, that the effects of a natural environment are almost entirely nullified unless the day's work contains certain elements that go to nourish a quiescent and receptive state of being. Arduous and incessant labour, even in the song-laden air of green fields, produces at the end of a day much the same sort of tired, introspective, unobservant man that it does in the grey street-crevices of a city. The ordinary work of a farm has so much exhausting bodily toil about it, as well as ceaseless repetition and dull routine, that, beyond a general interest in the weather as affecting his physical comfort or discomfort, the average farm hand has neither incentive nor inclination to look about him and cultivate an intellectual pleasure in wild natural things.

There are, however, two classes of country-dwellers who seem to represent the pure product of their environment; to be morally and physically made by, and for, the scenes and influences that surround them from earliest childhood. These, the woodlanders and the Southdown shepherds, form a very striking contrast to their fellows who labour in the fields. The gamekeepers must be set apart from the true woodland-folk, because their work is a continuous fight against natural conditions, and, as a general rule, they develop a correspondingly artificial habit of mind. But the wood-cutters, charcoal-burners, hurdle-makers, and the like are almost invariably men who take the deepest interest in the beautiful, strenuous life that surrounds them. It has always been difficult for one not of their class to get on

Shepherd and Flock 7

the right side of these solitary ruminating men ; and now that the wandering stranger in the by-ways is no longer a rarity, it is harder than ever to get on familiar terms with them ; but so soon as their inveterate shyness and taciturnity are charmed away, you are sure to find that you have sprung a veritable gold-mine of quaint and interesting information. One acquaintance formed in this way and adroitly cultivated, will prove of more value to the student of nature than all the books he could read in a decade. Yet the leafy solitude of the woodlands does not seem complete enough to give the tendencies of nature's own environment fullest and freest play. In the remotest places the worker is still in regular, if infrequent, touch with his fellows ; and the labour is almost as hard and unremitting as that of the open fields. In southern England, at least, the ideal conditions are to be found, perhaps, only in the life of the shepherd on the South Downs. To realise something of the conditions of his life, and what goes to the making of this gentle yet sturdy-natured philosopher of the wilds, it is necessary to be out and about on the Downs at all seasons, at all hours, and in every state of weather. A casual acquaintance with them is of little service. The chance wayfarer gets in fine weather only the impression of a vast, silent, sunny waste ; and in stormy times, of a desolation unkindly, almost terrible—something to flee from, as from the wrath to come. But though the shepherd's year contains its full share of vicissitudes and hardships, especially in winter and early spring, his life is in the main passed amidst thoroughly tranquillising and exhilarating influences.

You cannot live long in the Sussex Down-country without becoming aware of at least one unique quality about these breezy highland solitudes. A warm day on the South Downs is much the same joyous uplifting thing whether the month be January or June. In the lowlands every month writes its own unerring signature in the hedgerows and fields. But here there are no leafless boughs in winter, nor ruddy autumn foliage to mark the time of year. When spring is running like liquid gold through the grass of the valleys, on the Downs there is the same green billowing stretch of hill and dale under the same sunshine : the furze-brakes take on a little brighter sheen, and the misty coombes a deeper blue in the shadows. Summer comes, but with none of the broiling heat and riotous *carmagnole* of colour. The sun is higher and the shadows shorter ; the wild thyme purples the hill-side here and there ; the swifts skim the upland unceasingly ; but there is no inherent vital change. A sunny day on the Downs has little difference in winter or summer but of degree ; the seasons come and go ; yet there is always the green grass and the sunshine, the singing larks, the hovering sheep-bell music, and the wind that is never still.

Passing his whole day, and nearly every day of the year, in this intrinsically favourable environment, it is little wonder that the Southdown shepherd develops attributes of mind and character not to be met with in any other class of workers on the soil. But there is another, and a still more potent, agency at work. Almost any kind of strong-limbed humanity can be employed in the common labour of a farm, but the sheep-tender must be born to his craft. Heredity plays an all-important part in the making of a good shepherd ; it is hardly too sweeping an assertion to make if we say that there are no bad shepherds on the whole of the five hundred square miles' stretch of the South Downs.

Flock-masters are too wide-awake and wary a class to employ any but a capable and experienced man in work that stands at the very source of their prosperity. There are boys who take to shepherding from other walks of life, urged by a natural irresistible gift, and they do well at it. But it is essentially a family calling. Most shepherds have as long a pedigree behind them as the sheep themselves. The work is handed down from father to son, generation after generation, and there is a sort of family accumulation of skill and knowledge. The child is born within sound of the bleating of the flock. As soon as he is big enough he goes to help his father on the Downs, or in the lambing-yard. Presently he is made under-shepherd, puts on the yoke of responsibility inseparable from his kind, and by and by, when exposure to the rains and the rough north winds has at last conquered the sturdy physique of

By Habberton Lulham.

BORN TO THE CRAFT

his sire, the old man goes to the chimney-corner, and thence to his eternal pasturage; while the young man takes the ancient Pyecombe crook, which has been in his family for unnumbered ages, and steps out for the green sunny heights behind his travelling flocks, a fully-fledged master-shepherd. It is no uncommon thing to meet with men who themselves have forty or fifty years of this exacting, responsible, yet leisurely life behind them, and whose fathers and grandfathers had each also served their full half-centuries at the same Arcadian task.[1]

Solitude and an open-air life, combined with work affording much leisure for independent thought and little for idleness, would soon have a marked effect on any temperament, and the Southdown shepherd comes of a peculiarly susceptible race. But we are even now leaving out of the calculation a factor which is, in any right estimation of his chances, impossible to ignore. Most of us have climbed a hill on a fine Sunday morning, and, looking down on the city beneath us, have listened to the clamour of the church bells, softened and confused by the distance into one steady, pure, sweet note. There is something indescribably soothing and mentally stimulating in this far-off sound of bells in the sunshine that no ordered melody can bestow. Now, hardly any day in the year goes by when a like restful Sunday morning spirit of music is not abroad on the Sussex Downs. With the song of the larks overhead blent into a single pure cadence, and the incessant quiet tolling of the sheep-bells below, the shepherd's whole life moves to music of some kind or other. What, it may be asked, would happen to any of us if, instead of the grime and dust of a city, we could choose the clean, bright air of the countryside for our workaday environment? And what sort of man would the average Londoner become if he could exchange the deafening hubbub of Cheapside or Fleet Street for Southdown bell-music, with its sweet, dim echo of worship and rest?

[1] On a farm near Lewes, Sussex, the race of shepherds dates back to the time of Cromwell.—[*Author's Note*.]

CONTENTEDNESS OF SOUTHDOWN SHEPHERDS

By W. H. Hudson, 1900

One of the numerous, mostly minute, differences to be detected between the Downland shepherd and other peasants—differences due to the conditions of his life—refers to his disposition. He has a singularly placid mind, and is perfectly contented with his humble lot. In no other place have I been in England, even in the remotest villages and hamlets, where the rustics are not found to be more or less infected with the modern curse or virus of restlessness and dissatisfaction with their life. I have, first and last, conversed with a great many shepherds, from the lad whose shepherding has just begun, to the patriarch who has held a crook, and "twitched his mantle blue," in the old Corydon way, on these hills for upwards of sixty years, and in this respect have found them all very much of one mind. It is as if living alone with nature on these heights, breathing this pure atmosphere, the contagion had not reached them, or else that their blood was proof against such a malady.

One day I met a young shepherd in the highest part of the South Downs, who was about twenty-three years old, handsome, tall, well-formed, his face glowing with health and spirits. I shared my luncheon with him, and then sitting on the turf we talked for an hour about the birds and other wild creatures which he knew best. He told me that he was paid 12s. 6d. a week, and had no prospect of a rise, as the farmers in that part had made a firm stand against the high wages (in some cases amounting to 18s.) which were being paid in other parts. I was tempted as an experiment to speak slightingly of the shepherd's homely trade. It was all very well in summer, I said, but what about the winter, when the hills were all white with snow; when the wind blew so strong that a man could not walk against or face it; when it was wet all day, and when all nature was drowned in a dense fog, and you cannot see a sheep twenty yards off? "We are accustomed to all weathers," he replied ; "we do not mind the wet and cold—we don't feel it." I persisted that he earned too little, that shepherding was not good enough for him. He said that his father had been a shepherd all his life, and was now old and becoming infirm ; that he (the son) lived in the same cottage, and at odd times helped the old man with his flock, and was able to do a good many little things for him which he could not very well do for himself, and would not be able to pay a stranger to do. That, I said, was all right and proper ; but his father, being infirm, would not be able to follow a flock many years longer on the hills, and when the old man's shepherding days were over the son would be free. Besides, I added, a young man wants a wife— how could he marry on 12s. 6d. a week ? There came a pleasant far-away look in his eyes ; it could be seen that they were (metaphorically) turned inwards, and were occupied with the image of a particular, incomparable "She." He smiled, and appeared to think that it was not impossible to marry on 12s. 6d. a week !

To all who love the Sussex Downs and their people, it must be a source of regret that the old system of giving the shepherd an interest in the flock was ever changed. According to the old system he was paid a portion of his wages in kind—so many lambs at lambing-time ; and these, when grown, he was permitted to keep with the flock. At shearing-time he was paid for the wool, and he had the increase of his ewes to sell each year. He was thus in a small way in partnership with his master, the farmer, and regarded himself, and was regarded by others, as something more than a mere hireling, like the shepherd of to-day, who looks to receive a few pieces of silver at each week's end, and will be no better and no worse off whether the year be fat or lean.

One would imagine that the old system must have worked well on the Downs, as it undoubtedly does in other lands where I have known it, and I can only suppose that its discontinuance was the result of that widening of the line dividing employer from employed which has been so general. The farmer did not improve his position by the change. I believe he lost more than he gained ; it was simply that the old relations between master and servant were out of date. He was a better educated man, less simple in his life than his forefathers, and therefore at a greater distance from his shepherd ; it would remove all friction, and simplify things generally, to put the shepherd on the same level with the field labourer and other servants; and this was done by giving him a shilling more a week in exchange for the four or five or six lambs he had been accustomed to receive every year.

VASTNESS OF SOUTHDOWN SHEEP-WALKS

By Richard Jefferies, 1887

The shepherd came down the hill carrying his greatcoat slung at his back upon his crook, and balanced only by the long handle projecting in front. His dog was a cross with a collie : the old sheep-dogs were shaggier and darker ; most of the sheep-dogs now used were crossed with the collie, either with Scotch or French, and were very fast—too fast in some respects. He was careful not to send them much after the flock, especially after feeding, when, in his own words, "the sheep had best walk slow then, like folk"—like human beings, who are not to be hastened after a meal. If he wished his dog to fetch the flock, he pointed his arm in the direction he wished the dog to go, and said, "Put her back." Often it was to keep the sheep out of turnips or wheat, there being no fences. But he made it a practice to walk himself on the side where the care was needed, so as not to employ the dog unless necessary.[1]

There is something almost Australian in the expanse of Southdown sheep-walks, and in the number of the flocks, to those who have been accustomed to the small sheltered

---

[1] Ellis, in his *Shepherd's Sure Guide* (1749), remarks that a lame shepherd and a lazy dog are the best attendants on a flock of sheep, because they do not over-drive or worry them. But he also adds that a nimble shepherd and a nimble dog are best in an open country.—[*Author's Note.*]

meadows of the vales, where forty or fifty sheep are about the extent of the stock on many farms. The land, too, is rented at colonial prices, but a few shillings per acre—so different from the heavy meadow rents. But, then, the sheep-farmer has to occupy a certain proportion of arable land as well as pasture, and here his heavy losses mainly occur. There is nothing, in fact, in this country so carefully provided against as the possibility of an English farmer becoming wealthy. Much Downland is covered with furze, and some seems to produce a grass too coarse, so that the rent is really proportional. A sheep to an acre is roughly the allowance.

A Downland Sheep Fair

From all directions along the road the bleating flocks concentrate at the right time upon the hill-side where the sheep fair is held. You can go nowhere in the adjacent town except uphill, and it needs no hand-post to the fair to those who know a farmer when they see him, the stream of folk tend thither so plainly. It rains, as the shepherd said it would ; the houses keep off the drift somewhat in the town, but when the shelter is left behind, the sward of the hills seems among the clouds. The descending vapours close in the view on every side. The actual field underfoot, the actual site of the fair, is visible, but the surrounding valleys and the Downs beyond them are hidden with vast masses of grey mist. For a moment, perhaps, a portion may lift as the breeze drives it along, and the bold, sweeping curves of a distant hill appear; but immediately the rain falls again and the outline vanishes. The glance can only penetrate a few hundred yards ; all beyond that becomes indistinct, and some cattle standing higher up the hill are vague and shadowy. Like a dew, the thin rain deposits a layer of tiny globules on the coat ; the grass is white with them; hurdles, flakes, everything is as it were the eighth of an inch deep in water. Thus on the hill-side, surrounded by clouds, the fair seems isolated and afar off. A great cart-horse is being trotted out before the little

Shepherd and Flock   15

street of booths to make him show his paces. They flourish the first thing at hand—a pole with a red flag at the end—and the huge frightened animal plunges hither

SPRINGTIME IN THE SUSSEX WEALD
By Habberton Lulham.

and thither in clumsy terror. You must look out for yourself and keep an eye over your shoulder, except among the sheep-pens. There are thousands of sheep, all standing with their heads uphill. At the corner of each pen the shepherd plants his crook upright; some of them have long brown handles, and these are of hazel with the bark on; others are ash, and one of willow. At the corners, too, just outside, the dogs are chained, and in addition there is a whole row of dogs fastened to the tent-pegs. The majority of the dogs thus collected together from many miles of the Downs are either collies, or show a very decided trace of the collie.

One old shepherd, an ancient of the ancients, grey and bent, has spent so many years among his sheep that he has lost all notice and observation; there is no "speculation" in his eye for anything but his sheep. In his blue smock-frock, with his brown umbrella, which he has had no time or thought to open, he stands listening, all intent, to the conversation of the gentlemen who are examining his pens. He leads a young restless collie by a chain; the links are polished to a silvery brightness by continual motion; the collie cannot keep still—now he runs one side, now the other, bumping the old man, who is unconscious of everything but his sheep.

At the verge of the pens there stand four oxen with their yokes, and the long slender guiding-rod of hazel placed lightly across the necks of the two foremost. They are quite motionless, except their eyes, and the slender rod, so lightly laid across, will remain without falling. After traversing the whole field, if you return you will find them in exactly the same position. Some black cattle are scattered about on the high ground in the mist, which thickens beyond them and fills up the immense hollow of the valley.

In the street of booths there are the roundabouts, the swings, the rifle galleries—like shooting into the mouth of a great trumpet—the shows, the cakes and brown nuts and gingerbread, the ale barrels in a row, the rude forms and trestle tables; just the same, the very same we saw at our first fair five-and-twenty years ago, and a hundred miles away. It is just the same this year as last, like the ploughs and hurdles and the sheep themselves. There is nothing new to tempt the ploughboys' pennies—nothing fresh to stare at.

SHEPHERDS' HUTS ON THE SOUTH DOWNS

By M. A. Lower, 1854

Here are the very words of one who had himself carried the shepherd's crook and worn the shepherd's greatcoat for many years on these hills:—"The life of a shepherd in my young days was not the same as it is now. . . . You very seldom see a shepherd's hut on our hills in these times, but formerly every shepherd had one. Sometimes it was a sort of cave dug in the side of a bank or link, and had large stones inside. It was commonly lined with heath or straw. The part above ground was covered with sods of turf or heath, or straw, or boughs of hawth.[1] In rough, sluckish[2] weather the shepherd used to turn into his hut and lie by the hour together, only looking out once in a while to see that the sheep didn't stray away too far. Here he was safe and dry, however the storm might blow overhead, and he could sit and amuse himself as he liked best. If he could read, so much the better. It was in my hut, over in the next bottom to this, that I first read about Moses and his shepherding life, and about David's killing of the lion and the bear. Ah, how glad I felt that we hadn't such wild beasties to frighten and maybe kill our sheep and us. The worst we ever had to fear were the foxes that sometimes killed a young lamb or two. But there was otherwhile a crueller than that. If a ewe happened to get overturned on a lonesome part of the hill the ravens and carrion crows would come and pick out her eyes before she was dead. This happened to two or three of my ewes, and at last I got an old gun and shot all the crows and ravens I could get nigh. Once I shot an eagle, but that was the only eagle I ever saw. Since the hills have been more broken up by the plough, such birds are but seldom seen. There haven't been any wild turkeys either for many a year. I have heard my father say he killed two or three no great while before I was born. They used to call them bustards."

---

[1] Heath or gorse. Sussex.—[*Author's Note.*]
[2] Sluckish is not in Wright's *Dialect Dictionary*, but it appears to mean boggy or miry; cp. slock, a bog; slough, etc.—[*Author's Note.*]

A SHEPHERD'S BUSH

By WODEHOUSE R. H. GARLAND, 1910

Though it is repeated daily by thousands of lips, by travellers underground and above ground, by conductors, by booking clerks and porters—familiar over the continents of Europe, America, and Africa, and now in Far Eastern Japan as the home of universal " Exposition "—how many people comprehend the meaning of the term " Shepherd's Bush"? Still less, how many have found themselves ensconced within any such friendly shelter?

When this district was an open plain or heath affording pasturage for herds of cattle and flocks of sheep, doubtless there grew—not perhaps

> The hawthorn bush which stands beneath the shade,
> For talking age and whispering lovers made,

or that which Milton had in his mind when he wrote in *L' Allegro*—

> And every shepherd tells his tale
> Under the hawthorn in the dale ;—

but a strong, sturdy tree with well-worn boll, from which during time immemorial herds had kept watch and ward over their fleecy charges. There are in many parts of the country such shepherd's bushes still in existence, and as they are invariably situated on rolling downs and commons in out-of-the-way haunts of solitude, a short description may not be out of place. Imagine, then, a stiff thorn-bush with the sharp and slender leaf-stems starting about three feet from the ground. Instead of the bush being left to grow in the ordinary way, all the inner wood has been cut out until an oval cup has been formed by the sprouting outer branches growing densely together to a thickness of about eighteen inches, while the trunk of the bush forms at its top a platform within and a step without. On the top of this platform are placed and replaced, as required, bundles of clean wheaten straw, so that with a sack thrown over the inside of the cup to shield him from the prickles the shepherd can stand up with his arms resting on the edges of the bowl and look around him far and wide, watching the movements of his flock, knowing, as he does from long experience, that

> Indeed a sheep doth very often stray
> An' if the shepherd be awhile away.

When they are not thus occupied by their rightful tenants, the idler and the true holiday-maker who can find for himself one of these snug retreats, will desire nothing better than to recline on the straw with a sack at his back, reading and smoking throughout the livelong summer's day, though for his own comfort and for the preservation of an ancient landmark he must be careful not to set fire to the straw or the consequences will be, to say the least, uncomfortable. When such a bush is kept judiciously clipped and trimmed it forms an effective, artistic, and even beautiful piece of topiary work, especially in the months of May and early June, when the hawthorn is in full bloom.

JOHN DUDENEY

By W. H. Hudson, 1900

A shepherd of the South Downs, named John Dudeney, afterwards a schoolmaster in Lewes, where I believe one or two of his granddaughters still keep a school, was included by M. A. Lower in his *Worthies of Sussex*, on account of his passion for books and other virtues. And it will be allowed by every one that a poor peasant youth, who when shepherding on the hills acquired a knowledge of astronomy and of other out-of-the-way subjects, and taught himself to read the Bible in Hebrew, was deserving of a place among the lesser celebrities of his county.

The following is from John Dudeney's own account of his shepherding :—" I have sometimes been on the hills in winter from morning till night, and have not seen a single person during the whole day. In the snow I have walked to and fro under the shelter of a steep bank, or in a bottom or coombe, while my sheep have been by me, scraping away the snow with their forefeet to get at the grass, and I have taken my book out of my pocket, and as I walked to and fro in the snow have read to pass away the time. It is very cold on the Downs in such weather. I remember once, whilst with my father, the snow froze into ice on my eyelashes, and he breathed on my face to thaw it off. The Downs are very pleasant in summer.

"At midsummer 1799 I removed to Kingston, near Lewes, where I was under-shepherd for three years. My flock was large (1400), and my master, the head shepherd, being old and infirm, much of the labour devolved on me. While here I had better wages, £6 a year. I also had a part of the money obtained from wheatears, though we did not catch them here in great numbers—a dozen or two a day, seldom more. From what I have heard from old shepherds, it cannot be doubted that they were caught in much greater numbers a century ago than of late. I have heard them speak of an immense number being taken in one day by a shepherd of East Dean, near Beachy Head. I think they said he took nearly a hundred dozen— so many that he could not thread them on crow-quills in the usual manner, but took off his round frock (smock) and made a sack of it to pop them into, and his wife did the same with her petticoat. This must have been when there was a great flight. Their numbers now are so decreased that some shepherds do not set up any coops, as it does not pay for the trouble."[1]

A Sussex Shepherdess

I have described the sweetest musical voice heard in Sussex as that of a young girl in the Downs; another

---

[1] I regret that space does not admit of a longer account of this interesting shepherd, and that Stephen Blackmore must for the same reason be passed over with the following brief notice. In *Sussex Archæological Collections*, vol. xxxviii., 1892, is this note by C. J. P., which the Society kindly gave me permission to quote :—" The Society's Museum has received a large and interesting addition to the collection of prehistoric exhibits in the shape of some 700 Neolithic flint implements (Sussex) from the district of East Dean, comprising celts, hammer-stones, scrapers, flakes, chisels, etc., presented by Mr. Stephen Blackmore, a shepherd of East Dean, who, with a quick eye, much patience, and great discrimination, had by degrees accumulated a large and varied collection." For further particulars of this famous shepherd I refer my readers to *The Spirit of the Downs*, by Mr. Arthur Beckett (1909).—[*Author's Note.*]

Shepherd and Flock

Downland girl's voice was one of the acutest carrying voices I ever heard in my life. She was a-shepherding (a rare thing for a girl to do [1]) on the very high downs between Stanmer and Westmeston; and for two or three days during my rambles among the hills in that neighbourhood I constantly heard her oft-repeated calls and long piercing cries sounding wonderfully loud and distinct even at a distance of two miles or more away. It was like the shrill echoing cries of some clear-voiced big bird—some great forest fowl or eagle, or giant ibis, or rail, or courlan, in some far land where great birds with glorious voices had not all been extirpated. It was nice to hear, but it surprised me that all this outcry, heard over an area of seven or eight square miles, was necessary. At a distance of a mile I watched her, and saw that she had no dog, that her flock numbering nine hundred travelled a good deal, being much distressed by thirst, as all the dew-ponds in that part of the Downs were dry. When her far-sounding cries failed to make them turn, she had to go after them, and her activity and fleetness of foot were not less remarkable than her ringing voice; but I pitied her doing the work of a man and a dog as well in that burning weather, and towards evening on my way home I paid her a visit. She was rather a lean, wiry-looking girl, just over fifteen years old, with an eager, animated face, dark skin, and blackish fuzzy hair and dark eyes. She was glad to talk and explain it all, and had a high-pitched but singularly agreeable voice, and spoke rapidly and well. The shepherd had been called away and no shepherd-boy could be found to take his place; all the men were harvesting, and the flock had been given into her charge. The shepherd had left his dog, but he would not obey her; she had taken him out several days, leading him by a cord, but no sooner did she release him than he would run home, and so she had given up trying. I advised her to try again, and the next day I spent some time watching her, the dog at her

---

[1] John Dudeney remarked in 1849:—"My mother sometimes tended my father's flock while he went to sheep-shearing. I have known other shepherds' wives do the same; but the custom, like many others, is discontinued. I have not seen a woman with a flock for many years."—[*Author's Note.*]

side, calling and crying her loudest, and flying over the wild hill-side after the sheep; but the dog cared not where they went, and sullenly refused to obey her. Here is a dog, thought I, with good old-fashioned conservative ideas about the employment of women; he is not going to help them make themselves shepherdesses on the South Downs. A probably truer explanation of the animal's rebellious behaviour was given by a young shepherd of my acquaintance. The dog, he said, refused to do what he was told, because the girl was not his master's daughter, nor of his house. The sheep-dog's attachment to the family is always very strong, and he will gladly work for any member of it, but for no person outside. "My dog," he added, "will work as willingly and as well for any one of my sisters when I leave the flock to their care, as he will for me; but he will not stir a foot for any person, man or woman, not of the family." He said, too, that this was the common temper of the Sussex sheep-dog; faithful above all dogs to their own people, but suspicious of strangers, and likely at any time to bite the stranger's hand that caresses them.

I dare say he was right; I have made the acquaintance of some scores of these Downland dogs, and greatly admired them, especially their brown eyes, which are more eloquent and human in their expression than any other dogs' eyes known to me; yet it has frequently happened that, after I had established, as I imagined, a firm friendship with one, he has suddenly snapped or growled at me.

My account of that most extraordinary hullabaloo among the hills made by the young shepherdess has served to remind me of the subject I had set myself, and I have not yet touched upon—the wonderful silence of the Downs, and the effect of nature's more delicate music heard in such an atmosphere. That clear repeated call of the young shepherdess would have sounded quite different from six to eight hundred feet below on the flat weald, where it would have mixed with other sounds, and a denser atmosphere and hedges and trees would have muffled and made it seem tame and commonplace. On the great smooth hills, because of their silence and their thinner, purer atmosphere, it fell startlingly on the sense, and the prolonged cries had a wild and lonely expression.

A Picturesque Shepherd-Boy

During a walk among the South Downs one day in June, looking up from the valley I was in, I saw far up near the top of the hill in front of me a shepherd-boy standing motionless, his crook in his hand, his dog, held by a cord or chain, at his side. Wishing to have a talk with him, I began the ascent of the rather steep slope, and he, divining my intention, waited for me. As I came close to him he made a very pretty picture, standing against the blue sky, knee-deep in the tall grass, just beginning to flower, which covered that part of the Down. Among the grass sainfoin grew abundantly, and the green grass was sprinkled everywhere with the rose-red of its blossoming spikes. Even a very few flowers of any other colour would have taken something from the exquisite charm of that chance green and rose-red arrangement. But there were no other flowers. The young shepherd, aged about fifteen, had one of those perfectly Saxon faces which you see more in Sussex than anywhere else in England—a large round face, rosy brown in colour, shy blue eyes, and light brown hair, worn long. The expression, the shy yet pleased look—pleased that the monotony of his long solitary day should be broken by this chance encounter with a stranger—was childlike and very pretty. He wore loose-fitting grey clothes, and a round, grey peakless cap; and for ornament he had fastened in the middle of it, where there had perhaps once been a top-knot or ball, a big woolly thistle flower. It was really curious to note how that one big thorny flower-head with its purple disc harmonized with everything about the boy and gave him a strange distinction.

FROM SHEPHERD-LAD TO LAND-HOLDER

By J. Bateman, 1907

A well-known Sussex farmer relates some experiences of farm-life and of shepherding. He begins by saying that he believes that the over-amount of schooling in these days is the sole cause of the men leaving the land. He got his first job of work at fourpence a day, when about seven years old, as a shepherd-boy. Every morning he had to stand by an old farmer and remember the number of twenties as the sheep were counted, and make no mistake. His master, counting two at a time, said the words: — Onetherum, twotherum, cockerum, qutherum, setherum, shatherum, wineberry, wigtail, tarrydiddle, den (for twenty sheep). On one occasion the boy got confused, and the tally came wrong, which exasperated the old man. Holding a chain at full length, he swung it round, saying, " Dannelly, boy, 'tis aisy." The chain accidentally struck the lad on the nose, and he carries the mark to this day. He is also marked by the old reaping sickle. He watched the reaper sharpen his sickle and hang it in the hedge. Taking his chance, he secured the sickle, meaning to have a fine old time at reaping; on the first attempt he nearly severed his finger, and it now shows a scar and curvature. Later, when he grew to be a strong young man, and reaping had been superseded by fagging, he often, he says, fagged an acre of corn in a day. Our hero tells a tale of a mower. The man had to mow a field of peas. He brought with him in the morning a gallon jar of beer. The day grew hot, and the jar was soon emptied. The shepherd-boy was sent to get it refilled. This also disappeared. Next morning the mower, on surveying his previous day's work, found that it only measured a few feet each way. " Well, boy," he said ruefully, " I shall have to work on water to-day." The jar was filled with water, and the work was finished that day. Once at shearing-time the over-generous sheep-shearers gave him more than was good for him of their cider allowance. The boy failed to come home at bedtime. Alarm was felt, and the old barn and surroundings were searched, and the pond near

"SOUTHDOWNS"
"Models of what hill sheep should be."
By Habberton Lulham.

was dragged, but without avail. In the morning, when the shearers arrived, no little merriment was caused on seeing him creep out from the folds of a pack of newly shorn wool. It is a remarkable fact that the shepherd-boy of those days now farms an area of considerably over two thousand acres, the whole, or nearly the whole, of the farm-land in a Sussex village.[1]

A HAMPSHIRE SHEPHERD

By H. Rider Haggard, 1902

Upon this farm, as on the majority in this neighbourhood, sheep of the Hampshire Down breed are the mainstay.

The shepherd, George Piper by name, was a person worthy of remark. Then over seventy years of age, for sixty of them he had been a shepherd, forty years of that time being spent in the employ of a single master. Sunk as he was in eld, it was easy for any one accustomed to watch his class to see that in very many cases the services of two men of the present generation would be of less value than those of this shepherd, who knew his sheep and was known of them.[2]

There he stood in the cold wind, upon the bleak down crest, watching the fold much as a dog does, and now and again passing the hurdles to do some little service to his flock, every one of which he could distinguish from the other. This, too, on a Sunday, the day on which it is so difficult to keep the modern stockman to work, however necessary.

The view from the sheep-fold was very fine. On the south, wide, open country, running to the dim line of the old Danish encampment of Danebury Ring; on the southwest, the dense mass of Wherwell Woods, still tinged with their autumn foliage.

---

[1] He is now (1910) the *owner* of a farm in a neighbouring parish.
[2] The good old shepherds don't think much of some of the young ones, and call them hurdle-pitchers. Shepherd Aylward of Lavant, when past work, would say, in bad weather, "Now I wonder how my sheep are faring; that young one won't be looking after them. I wish I could go and see to them." In the past he had never grudged any time or trouble, day or night, spent for the welfare and happiness of his dear flock.—[*Author's Note.*]

DORSETSHIRE

HARDINESS OF THE PORTLAND BREED

By The Author

Portland's "abrupt peninsula of rock," with its steep sides and flat summit, was in old days a well-known sheep-run, celebrated for its breed of sheep. Portland sheep are now but few, and number about one hundred. Some Southdown sheep are kept, and these probably number from three to four hundred. Portland sheep are very small — smaller than Southdowns. They have long woolly tails, brown faces, knees, and feet; their horns are flattened, with slight indentations along them, marked with wavy lines. They are a breed peculiar to the island, where they thrive well, and are easy to rear, and farmers are now endeavouring to increase their stock. A number have been imported into the Channel Islands, where they are highly valued. Their chief characteristics are their hardiness, and the fact that they make excellent mutton. The breed has been dying out in Portland, probably because the more profitable trade of stone-quarrying has in many cases superseded grazing. The grazing is generally flat, but the sheep are but roughly attended to; they are, however, folded at night. The English sheep-dog is used; the Portland sheep are said to resent the collie, and butt him. In days gone by the Portland farmer was, as a rule, his own shepherd, but girls also were employed for sheep-tending. Then, as now, holdings were small, and there were no large farms, but a considerable quantity of common and open lands. The farmers kept to the same pure Portland breed and the same flock.

DEVONSHIRE

COURAGE OF THE EXMOOR SHEEP

By J. H. Crabtree, 1909

I have noticed the courageous bearing of sheep on the hills about Exmoor. Here the animals roam over miles of uncultivated land, and for days see not a living soul. If suddenly disturbed from some unexpected quarter they do not take flight, but toss their heads and spring forward side by side, as if forming a close marching line, or an impregnable rampart. In the lambing season this innate courage is exhibited most strongly. If stray dogs are near the sheep-fields, the mothers will collect their little snow-white lambs and shield them resolutely from impending danger. The dogs may bark and snarl, but woe betide the canine invader who ventures to within ten inches of a ewe's horns.[1] It is true that some sheep do not attain to this defensive degree, but even these will shield their progeny to the last and will not allow the lambs to be worried. The sheep is an admirable creature.

CORNWALL

SHEEP OF THE "TOWENS" AND SCILLY

By William Borlase, F.R.S., 1758

In the neighbourhood of St. Columb and St. Kevern the sheep are large, and bring a great price, but the sweetest mutton is reckoned to be that of the smallest

---

[1] Hogg, in his *Mountain Bard*, has : " However excited and fierce a ewe may be, she will never offer any resistance to mankind, being perfectly and meekly passive to them." But Ralph Fleesh, whose valuable opinion on this matter differs from Hogg's, writes : " The female never attacks a man unless by way of guarding her lambs. When her lamb is only a day or two old, and she is approached by a stranger, her attitude is sometimes strongly defiant. The male also, particularly when hand-fed, will attack a man at close quarters."—[*Author's Note.*]

sheep, which usually feed on the commons where the sands are scarce covered with the green sod, and the grass exceedingly short; such are the towens or sand-hillocks in Piran-sand, Gwythien, Philac, and Senangreen near the Land's End, and elsewhere in like situations. From these sands come forth snails of the turbanated kind, but of different species and all sizes, from the adult to the smallest just from the egg. These spread themselves over the plains early in the morning, and whilst they are in quest of their own food among the dews yield a most fattening nourishment to the sheep.[1]

In the ISLES OF SCILLY[2] sheep thrive exceedingly,[3] the grass on these commons being short and dry, and full of the same little snail which gives so good a relish to the Senan and Philac mutton in the west of Cornwall. The sheep will fill themselves upon the ore-weed,[4] as well as the bullocks. They have sea-wrack among their ore-weed of a fine scarlet and other colours, and good laver.[5]

SOMERSETSHIRE

BEN BOND, *IDLETON*

By JAMES JENNINGS, 1834.

Agricultural and other labourers, whose exertions are usually great, and sometimes from long continuance extremely grievous, are apt to suppose that the chief happiness of life consists in having *nothing to do*; in a

[1] Cp. "Southdowns and Snails," *infra*, p. 116, where the eating of snails by pasturing sheep is noted and explained.
[2] Borlase's *Islands of Scilly*, 1756.
[3] There are but few sheep in the Isles of Scilly now, and these are described as of "rather a scraggy order." As there are no regular shepherds the sheep are spanned to avoid incursion into neighbouring fields. As much of the land is now devoted to "flower-farming," it is a good thing that so many snails were consumed in 1756!—[*Author's Note.*]
[4] Ore (Saxon), seaweed washed ashore. See Cooper's *Glossary of Provincialisms in use in the County of Sussex*, 1853. The proper spelling is "woar."—[*Author's Note.*]
[5] Laver, a kind of seaweed from which the well-known Devonshire dish is prepared. —[*Author's Note.*]

word, in being *idle*. Alas! how mistakenly ignorant are such persons of their own nature.

Ben Bond was one of those sons of idleness whom ignorance and want of occupation in a secluded country village too often produce. He was a comely lad, on the confines of sixteen, employed by Farmer Tidball, a querulous and suspicious old man, to look after a large flock of sheep. The scene of his soliloquy may be thus described. A green sunny bank, on which the body may agreeably repose, called the *Sea Wall*. On the sea side was an extensive common called the *Wath*, and adjoining to it was another called the *Island*; both were occasionally overflowed by the tide. On the other side of the bank were rich enclosed pastures, suitable for fattening the finest cattle. Into these enclosures many of *Ben Bond's* charge were frequently disposed to stray. The season was June, the time midday, and there will be no anachronism in stating that the western breezes came over the sea a short distance from which our scene lay, at once cool, grateful, refreshing, and playful. The rushing *Parret*, with its ever-shifting sands, was also heard in the distance. It should be stated, too, that *Larence* is the name usually given in Somersetshire to that imaginary being which presides over the *Idle*.[1] Perhaps it may also be useful to state here that the word *Idleton*, which does not occur in our dictionaries, is assuredly more than a provincialism, and should be in those definitive assistants.

During the latter part of the soliloquy Farmer Tidball arrives behind the bank, and hearing poor Ben's discourse with himself, interrupts his musings in the manner described hereafter. Whether it be any recommendation to this soliloquy or not, the reader is assured that it is the history of an occurrence in real life, and that it happened at the

---

[1] As lazy as Laurence (St. Laurence, who suffered martyrdom about the middle of the third century). *Notes and Queries* gives the following :—"That famous and resolute champion Laurence, who was roasted on a grid-iron." A tradition has been handed down from age to age that at his execution he bore his torments without a writhe or groan, which caused some of those standing by to remark, "How great must be his faith!" But his pagan executioner said, "It is not his faith, but his idleness; he is too lazy to turn himself." And hence arose the saying, "As lazy as Laurence."— [*Author's Note.*]

place mentioned. The writer knew Farmer Tidball personally, and has often heard the story from his wife.

SOLILOQUY

"Larence! why doos'n let I up? Oot let I up?"—"Naw, I be slĕapid; I can't let thee up eet." "Now, Larence! do let I up. There! bimeby maester 'll come, an a'll beät I athin a ninch[1] o' me life; do let I up!"—"Naw, I wunt."

"Larence! I bag o' ee, do ee, do ee let I up. D'ye zee tha sheep be all a-breakin' droo tha hadge inta tha vive-an-twenty yacres, an Former Haggit 'll goo ta La wi'n, an I sholl be a kill'd!"—"Naw, I wunt—'tis zaw whit; bezides, I hant a had my nap out."

"Larence! I da za thee bist a bad un. Oot thee hire what I da za? Come now an' let I scoose[2] wi'. Massy on me! Larence, whys'n thee let I up?"—"Câz I wunt. What, muss'n I ha an hour like uither vawk ta ate my bird an cheese? I do zá I wun't; and zaw 'tis niver-tha-near[3] to keep on."

"Maester twal'd I, nif I war a good bway, a'd gee I iz awld wâsket; an' I'm shower, nif a da come an' vine I here, an' tha sheep a brawk inta tha vive-an'-twenty yacres, a'll vleng't awâ vust! Larence, do ee, do ee let me up! Ool ee, do ee!"—"Naw, I tell ee I wunt."

"There's one o' tha sheep 'pon iz back in tha gripe,[4] an a can't turn auver! I mis g'in ta tha groun an g'out to'n, an' git'n out. There's another in tha ditch! a'll be a buddled![5] There's a gird'l[6] o' trouble wi' sheep! Larence, cass'n thee let I goo? I'll gee thee a ha'peny nif oot let me."—"Naw, I can't let thee goo eet."

"Maester 'll be shower to come an' catch me! Larence! doose thee hire? I da za, oot let me up. I zeed Forerm

---

[1] Ninch = inch. Such instances of wrong division, where the *n* of the indefinite article precedes a vowel, are common. [2] Scoose wi' = discourse or talk with you. [3] Niver-tha-near, *adv.* to no purpose, uselessly; near here stands for nearer.—[*Author's Note.*]
[4] Gripe, *s.* a small drain or ditch, about a foot deep and six or eight inches wide.
[5] Buddled, *part.* suffocated in mud.
[6] Gird'l, *s.* contracted from great deal and implying the same; as gird'l o' work, a great deal of work.

Haggit zoon ater I upt, an' a zed, nif a voun one o' my sheep in tha vive-an'-twenty yacres, a'd drash I za long as a cood ston auver me, an' wi' a groun ash, too! There! zum o'm be a gwon droo tha vive-an'-twenty yacres inta tha drauve;[1] thâ'll zoon hirn[2] vur anow. Thâ'll be poun'd.[3] Larence, I'll gee thee a penny nif oot let me up."—" Naw, I wunt."

"Thic not-sheep[4] ha' got tha shab! Dame tawl'd I whun I upt ta-da ta mine tha shab-water;[5] I sholl pick it in whun I da goo whim.[6] I vorgoot it! Maester war desperd cross, an' I war glad ta git out o' tha langth o' iz tongue. I da hate zitch cross vawk. Larence! what, oot niver let I up? There, zum o' tha sheep be agwon down ta Ready Ham; withers be gwon into leek-beds; an' zum o'm be in Hounlake; dree or vour o'm be gwon za vur as Slow-wá; the ditches be, menny o'm, zâ[7] dry 'tis all now rangel common. There! I'll gee thee dree hā pence ta let I goo."—" Why, thee hass'n bin here an hour, an' vor what shood I let thee goo? I da za, lie still!"

"Larence! why doos'n let I up? There! zim ta I, I da hire thic pirty maid, Fanny o' Primmer Hill, a chidin bin I be a lyin' here while tha sheep be gwain droo thic shord an' tuther shord;[8] zum o'm, a-ma-be, be a drown'd. Larence! doose thee thenk I can bear tha betwitten o' thic pirty maid? She tha Primrawse o' Primmer Hill; tha lily o' tha level; tha gawl-cup[9] o' tha mead; tha zweetist honeyzuckle in tha garden; tha yaly vilet; the rawse o' rawses; tha pirty pollyantice;[10] whun I zeed 'er last, she zed, 'Ben, do ee mind tha sheep, an' tha yeos an' lams, an' than zumbody ool mine you.' Wi' that she gid me a beautiful spreg o' jessamy, jist a-pickt vrom tha poorch,— tha smill far za zweet.

"Larence, I mus'[11] goo! I ool goo. You mus let I

---

[1] Drauve = a drove, or road to fields. [2] Hirn, v.n. to run.
[3] "To poun," and not "to pound," is the verb in Somersetshire.
[4] Not-sheep, s. sheep without horns.
[5] Water to cure the shab or itch in sheep.
[6] Whim, s. home. [7] Zâ = say.
[8] Shord, s. a sherd, a gap in a hedge. Stop-sherd, a stop-gap.
[9] Gawl-cup, s. gold-cup.
[10] Pollyantice, s. polyanthus.
[11] Mus' goo = must go. This dropping of the final t is by no means uncommon.

Shepherd and Flock 33

up. I ont stâ here na longer. Maester 'll be shower ta come an' drash me. Thic awld cross fella wi' iz awld wåskit. There, Larence! I'll gee tuther penny, an' that's ivry vard'n I 'a got. Oot let I goo?"—"*Naw, I mis ha a penny moor.*"

"Larence, do let I up! Creeplin Philip 'll be shower ta catch me! Thic cockygee![1] I dwont like en at all; a's za rough an za zoür. An Will Popham too, ta bewite me about tha maid! A câll'd 'er a rathe-ripe Lady Buddick.[2] I dwont mislike tha name at âll, thawf I dwont care vor'n a stra, nor a read mooäte, nor tha tite o' a pin! What da tha câll he? Why, the upright man, câs a da ston upright ; let'n ; and let'n wrassly[3] too. I dwont like zitch hoss-plâs,[4] nor singel-stick nuther, nor cock-squailin',[5] nor menny wither mâ-games that Will Popham da volly. I'd rather zit in tha poorch, wi' tha jessamy ranglin roun it, and hire Fanny zeng. Oot let me up, Larence?"—"*Naw, I tell thee I ont athout a penny moor.*"

"Rawzy Pink, too, an Nanny Dubby axed I about Fanny. What bisniss 'ad thâ ta up wi't? I dwont like norn o'm. Girnin Jan, too, shawed iz teeth an put in iz verdi. I wish theeäze vawk ood mine ther awn consarns an let I an Fanny alooäne.

"Larence! doose thee meän to let I goo?"—"*Eese, nif thee't gee me tuther penny.*" "Why, I hant a-got a vard'n moor ; oot let me up!"—"*Not athout tha penny.*" "Now, Larence! de ee, bin I hant naw moor money. I a bin here moor than an hoür ; whaur tha yeos an lams an âll tha tuther my sheep be now I dwon'[6] know. Creeplin Philip[7] ool gee me a lirropin shower anow. There!

[1] Cockygee, *s.* cockagee, a rough, sour apple.
[2] Lady Buddick, a rich and early ripe apple. Rathe-ripe, *adj.* ripening early. "The rathe-ripe wits prevent their own perfection."—BISHOP HALL.
[3] Wrassly = wrestle.
[4] Hoss-plâs, *s.p.* horse plays ; rough sports.
[5] Cock-squailin', a barbarous sport, consisting in tying a cock to a stake, and throwing a stick at him from a given distance, until the bird is killed. It was once common at Shrovetide.
[6] Here, instead of *don't*, or *dwont*, for *do not*, we have *dwon'* only, which, in colloquial language, is very common in the west.
[7] Even remote districts in the country have their satirists, and wits and would-be wits ; and Huntspill, the place alluded to in the soliloquy, was, about half a century ago,

D

I da thenk I hired zummet or zumbody auver tha wall."

" Here, d——n thee! I'll gee tha *tuther penny, an zummet bezides!*" exclaimed Farmer Tidball, leaping down the bank, with a stout sliver of a crab-tree in his hand. The sequel may be easily imagined!

WILTSHIRE

THE SHEPHERD OF THE PLAIN

By PERCY W. D. IZZARD, 1910

Take the year round, and Salisbury Plain sees far more of soldiers and sheep than of any other living creatures. The soldiers are of comparatively recent advent, and come only in large numbers at certain times of the year; but the sheep are always there in great flocks, and they have roamed the solitudes of the plain for countless generations.

These remote and ample Downs are an ideal ground for the practice both of warlike arts and pastoral pursuits, but it was to see a shepherd rather than a soldier that I went upon the Plain on a sunny winter's day. From an eminence one could see in the clear light almost the entire extent of the great airy Downland, and flocks of sheep on every side eating their way hither and thither beneath the blue sky. These were flocks of the farms that fringe the plain.

At one of the farms I found my shepherd, and he was different from all other shepherds I know. One of many brothers, he had to choose a calling, and he decided

---

much pestered with them. Scarcely a person of any note escaped a parish libel, and even servants were not excepted. For instance—

" Nanny Dubby, Sally Clink,
Long Josias and Rawsy Pink,
. . Girnin Jan,
Creeplin Philip and the Upright Man."

Creeplin Philip (that is "creeplin," because he walked lamely) was Farmer Tidball himself; and his servant, William Topham, was the *upright man.* Girnin Jan is Grinning John.

Shepherd and Flock 35

to be a flock-master. He loves his profession, and there is little that he can be taught about sheep. He also loves a bit of sport—shooting or hunting—and before he came away from his native Devon he often rode to hounds.

He is comparatively a young man, of middle height and sturdy build, with clean-shaven chin, crisp moustache, and generally enlightened face. His cloth cap, riding-breeches and gaiters complete the illusion that he is not a shepherd ; but his success in his calling has gained him fame beyond the Plain's borders. His whole appearance suggests great strength and staying power, and those he has. . . .

The ewe flock with some of the earlier lambs were out on the Plain, 2000 strong and widely dispersed. That explained my shepherd's costume, for he was compelled to get about a great deal on horseback. In miniature his methods are those of a large colonial sheep-farmer, and about 5000 acres of the green Downs are rented for the sheep. He would not have come up here to Salisbury Plain if there had been no Exmoor sheep to look after. It is the breed of his own country, and he knows and understands it best of all. About the Plain other shepherds have charge of Hampshire Downs, Southdowns, Scotch, and Cheviot sheep, and other methods prevail than those of this strong-framed Devonian.

. . . .

It is the last hour of this crystal winter's day, whose embers glow red-gold across the western ridges. The frost is returning, stiffening again the mud in the yard. A storm-cock in the trees hard by is singing the world to rest. The shepherd mounts his horse and rides forth upon the plain, with his dog Bolt, a black shaggy fellow of the old English type, in vigilant attendance. Up and down he rides and up again, ever rising, man and horse diminishing in size every moment, until a stop is made on a commanding summit, where they are silhouetted against the rosy afterglow. There the shepherd can see his widely scattered flock before him over the far-spread Downs. Then he lifts up his great voice—" Yaa-hoo ! Yaa-hoo!" and once more, "Yaa-hoo!"[1] The call carries through the motionless air a mile and more into the silence of the plain. "Bolt, get up for 'em," and off scuttles that shaggy servant. But the sheep know the call and the order of things. You can see those grey dots, however far away, begin to converge slowly upon one centre. The sheep are forming up their flock themselves, and Bolt's duty is but to hasten them. Very soon the dog is little more than a speck in the distance, racing round and round the sheep until the latter begin to move forward in one great body—a grey cloud coming hither over the darkling green. Presently the voice of the flock is heard. Nearer, louder, nearer still and louder, and at last the woolly company jostles with many deep "baaings" down the last slope to the sheltering trees, horseman and dog following slowly behind. Once in the hollow where the night is to be spent, the matronly ewes soon settle. In the meantime, in the deepening twilight, shepherd and dog have turned homeward—the day's work done.

LAZY SHEPHERDS OF THE PLAIN—AND AN EXCEPTION

By THE AUTHOR

The shepherds of Salisbury Plain had the reputation of being the laziest men in England. A gentleman saw one of them lying down near his sheep, and after talking with him for some time, said: "Well, my man, I do think you are as lazy as you are described; nevertheless here's a shilling for you." To which the man replied: "Thank ye kindly, sir, but will ye jist get off your horse and slip it in my pocket."

In Hone's *Every-Day Book*, for July 1826, we find: "The shepherds of Salisbury Plain are proverbially so idle, that rather than rise, when asked the road across the plain, they put up one of their legs towards the place, and say, '*Theek woy*' (this way), '*Thuck woy*' (that way)."

[1] "Shouting as loud as if he was calling sheep on Exmoor." See Blackmore's *Lorna Doone*.—[*Author's Note*.]

Shepherd and Flock

Hannah More's exemplary *Shepherd of Salisbury Plain* was a great exception to these lazy fellows. I give in full the note to her little book referring to him. " David Saunders was a poor shepherd of West Lavington. He used to keep his Bible in the thatch of his hut on Salisbury Plain, by reading which, and by prayer, he seemed to keep up a constant communion with God. When the late Mr. Stedman, of Shrewsbury, went in 1771, to settle on the curacy at Little Cheverell, the next village to Lavington, the first person he met was this shepherd, who told him, some time after, that, taking the stranger to be the minister expected there, he repeated to himself those words of St. Paul, Romans x. 15 : 'How beautiful upon the mountains are the feet of them that preach the gospel of peace, and bring glad tidings of good things.'"

Dr. Brewer tells us that David Saunders and his father before him kept sheep on the Plain for a century, and that David was noted for his homely wisdom and simple Christian piety.

WILTSHIRE SHEPHERDS, 17TH CENTURY

From *The Book of Days.* Edited by R. CHAMBERS, 1869

John Aubrey was a native of Wiltshire, and therefore proud of its Downs, which, in his odd, quaint way, he tells us, " are the most spacious plains in Europe, and the greatest remains that I can hear of the smooth primitive world when it lay under water. The turf is of soft sweet grass, good for the sheep. About Wilton and Chalke, the Downs are intermixt with boscages,[1] that nothing can be more pleasant, and in the summer-time do excel Arcadia in verdant and rich turf." Then, pursuing the image, he says : " The innocent lives of the shepherds here do give us a resemblance of the Golden Age. Jacob and Esau were shepherds ; and Amos, one of the royal family, asserts the same of himself, for he was among the shepherds

---

[1] Woodlands. According to Blount, "that food which wood and trees yield to cattle."—HALLIWELL.—[*Author's Note.*]

of Tecua (Tekoa), following that employment. The like labour, by God's own appointment, prepared Moses for a sceptre, as Philo intimates in his *Life*, when he tells us that a shepherd's art is suitable preparation to a kingdom. The same he mentions in his *Life of Joseph*, affirming that the care a shepherd has over his cattle very much resembles that which a king has over his subjects. The same St. Basil, in his homily *De St. Mamene, Martyre*, has concerning David, who was taken from following the ewes great with young ones to feed Israel. The Romans, the worthiest and the greatest nation in the world, sprang from shepherds. The augury of the twelve vultures placed a sceptre in Romulus's hand, which held a crook before, and as Ovid says :

> His own small flock each senator did keep.

Lucretius mentions an extraordinary happiness, and as it were divinity, in a shepherd's life :

> Thro' shepherds' care, and their divine retreats.

And to speak from the very bottom of my heart, not to mention the integrity and innocence of shepherds, upon which so many have insisted and copiously declaimed, methinks he is much more happy in a wood, that at ease contemplates the universe as his own, and in it the sun and stars, the pleasing meadows, shaded groves, green banks, stately trees, flowing springs, and the wanton windings of a river, fit objects for quiet innocence, than he that with fire and sword disturbs the world, and measures his possessions by the waste that lies about him."

Then the old Wiltshire man tells us how the Plains abound with hares, fallow-deer, partridges, and bustards. The fallow-deer and bustards have disappeared. In this delightful part of the country is the Arcadia about Wilton which " did no doubt conduce to the heightening of Sir Philip Sydney's phansie. He lived much in these parts, and the most masterly touches of his pastorals he wrote here upon the spot, where they were conceived. 'Twas about these purlieus that the Muses were wont to appear to Sir Philip Sydney, and where he wrote down their dictates in his table-book though on horseback," and some old relations of Aubrey's remembered to have seen Sir Philip do this.

Aubrey then proceeds to trace many of the shepherds' customs of his district to the Romans, from whom the Britons received their knowledge of agriculture. The festivals at sheep-shearing he derives from the Palilia. In

By Habberton Lulham.

A SHEPHERD'S CARE

Aubrey's time, the Wiltshire sheep-masters gave no wages to their shepherds, but they had the keeping of so many sheep *pro rata*, "so that the shepherds' lambs do never miscarry"; and Plautus gives a hint of this custom amongst the Romans at his time.[1] In Scotland it is still the custom to pay shepherds partly in this manner. The Wiltshire antiquary goes so far as to say that the habit of his time was that of the Roman, or Arcadian shepherds,

[1] John Aubrey, in *Old Manners and Customs* (1678), says: "In our west country (and I believe so in the north) they give no wages to the shepherd, but he has the keeping so many sheep with his master's flock. Plautus hints at this in his *Asinaria*, act iii. scene 1. *etiam Opilio.*"—[*Author's Note.*]

as delineated by Drayton in his *Polyolbion, i.e.* a long white cloak with a very deep cape, which comes half-way down their backs, made of the locks of the sheep. There was a sheep-crook, as we read of in Virgil and Theocritus ; their sling, scrip, tar-box, a pipe or flute, and their dogs.[1]

WILTSHIRE SHEPHERD CUSTOMS

By Richard Jefferies, 1880

Some of the older shepherds still wear the ancient blue smock-frock, crossed with white " facings," like coarse lace ; but the rising generation use the greatcoat of modern make, at which their forefathers would have laughed as utterly useless in the rain-storms that blow across the open hills. Among the elder men, too, may be found a few of the huge umbrellas of a former age, which when spread give as much shelter as a small tent. It is curious that they so rarely use an umbrella in the field, even when simply standing about ; but if they go a short journey along the highway, then they take it with them. The aged men sling these great umbrellas over the shoulder with a piece of tar cord, just as a soldier slings his musket, and so have both hands free—one to stump along with a stout stick, and the other to carry a flag basket. The stick is always too lengthy to walk with as men use it in cities, carrying it by the knob or handle ; it is a staff rather than a stick, the upper end projecting six or eight inches above the hand.

If any labourers deserve to be paid well, it is the shepherds ; upon their knowledge and fidelity the principal profit of a whole season depends on so many farms. The shepherd has a distinct individuality, and is generally a much more observant man in his own sphere than the ordinary labourer. He knows every single field in the whole parish, what kind of weather best suits its soil, and can tell you without going within sight of a given farm pretty much what condition it may be found in. Know-

[1] Aubrey's *Natural History of Wilts.*

Shepherd and Flock 41

ledge of this character may seem trivial to those whose

MOTHERING

"It needs a man who has grown gentle and almost motherly to be a shepherd, there is so much mothering to be done."—H. SOMERSET BULLOCK.

"A hard-hearted shepherd is, of all problems, the most inexplicable and painful."—JAMES GARDNER.

days are passed indoors; yet it is something to recollect all the endless fields in several square miles of country.

As a student remembers for years the type and paper, the breadth of the margin—can see, as it were, before his eyes the bevel of the binding and hear again the rustle of the stiff leaves of some tall volume which he found in a forgotten corner of a library, and bent over with such delight, heedless of dust and "silver-fish" and gathered odour of years—so the shepherd recalls *his* books, the fields; for he, in the nature of things, has to linger over them and study every letter: sheep are slow.

When the hedges are grubbed up, and the grass grows where the hawthorn flowered, still the shepherd can point out to you where the trees stood—here an oak and there an ash. On the hills he has often little to do but ponder deeply, sitting on the turf of the slope, while the sheep graze in the hollow, waiting for hours as they eat their way. Therefore, by degrees, a habit of observation grows upon him—always in reference to his charge; and if he walks across the parish off duty he still cannot choose but notice how the crops are coming on, and where there is most "keep." The shepherd has been the last to abandon the old custom of long service. While the labourers are restless, there may still be found not a few instances of shepherds whose whole lives have been spent upon one farm. Thus, from the habit of observation and the lapse of years, they often become local authorities; and when a dispute of boundaries or water rights or right of way arises, the question is frequently finally decided by the evidence of such a man.

Every now and then a difficulty happens in reference to the old green lanes and bridle-tracks which once crossed the country in every direction, but get fewer in number year by year. Sometimes it is desired to enclose a section of such a track to round off an estate—sometimes a path has grown into a valuable thoroughfare through increase of population; and then the question comes, Who is to repair it? There is little or no documentary evidence to be found — nothing can be traced except through the memories of men; and so they come to the old shepherd, who has been stationary all his life, and remembers the condition of the lane fifty years since. He always liked to drive his sheep along it—first, because it saved the turnpike tolls; secondly, because they could graze on the short herbage and rest under the shade of the thick bushes. Even in the helplessness of his old age he is not without his use at the very last, and his word settles the matter.

CHARACTERISTIC WILTSHIRE SHEPHERDS

By A. G. Bradley, 1907

A great deal of precious immemorial turf was ploughed up on Salisbury Plain when wheat was high, to grow sometimes but second-rate crops of three quarters to an acre, and many a bitter regret has doubtless been since felt for the destruction of sheep-pasture that takes nearly half a century to regain its original quality. The military quarter of the Plain beyond the Avon, from which the sheep have been banished, and the down grass, for centuries cropped close by them, which has been allowed to seed, or casually picked at by cows and horses, presents a deplorable sight, though of comparatively slight moment, in view of the much worse disfigurement of the military buildings, where such conditions prevail. But here, on the north and west portion of the Plain, we are far away from all such desecrations, and the sheep still keeps the succulent turf, mixed with nutritious plants and flowers of great variety, sweet and short as he has kept it, no doubt, since his breed was first introduced into England.

It is thought that the great Wiltshire sheep fairs are the oldest fixtures of the kind in England; and they have a double interest, from being, in some cases, held by immemorial custom on lonely points like Tan Hill, marked by a prehistoric camp. Thousands of sheep collect on these occasions, travelling long journeys on foot, cropping their slow way along the broad green trails on the high down followed for centuries by Wiltshire shepherds, whose rugged figures and simple equipments have not perhaps changed much more in appearance than the bare hills around them.[1] I came across a fine figure of a Wiltshire shepherd one morning, standing out against the sky on the rampart of a British camp near Ogbourne, overlooking half the county. He was a quite ideal picture, his crook in his hand, his cloak beside him on the ground, and lying upon it a shaggy grey sheep-dog, which eyed me suspiciously, while the steady crunching all around of what Thomson would have called his "fleecy care" was the only audible sound upon the waste. He most assuredly was not posing, for his condition, I found, was one of despair. "I've had a turr'ble misfart'n, zur, this marnin'" (he might have said "this marnin's marnin'," for that is good old Wiltshire); "I larst my bacca pouch." An unworthy distrust of human nature prompted the thought that he was as other degenerate mortals, and the picture for the moment lost some value. But I had a pleasant surprise. "No, zur, thank 'ee kindly. 'Tain't the bacca, but the pouch; ma zon guv it I my larst birthday. It's a turr'ble misfart'n I should ha' larst'n." He would neither accept of nor be comforted from my supplies, and after he had dwelt on the changes that had come over the country, in critical and pessimistic fashion, and with no apparent thankfulness for the rise of sixty per cent in wages, I left him still unconsoled. . . . Silbury is as grim and naked as in that remote age when the turf first clothed it. Together with its encircling ditch, it covers an area of five acres, and is 170 feet high. Though Charles II. and his brother James, it will be

---

[1] Mr. Walter Money, F.S.A., in a letter to me (1910) writes: "The drovers of the flocks from the great sheep fairs at Weyhill (Hants), Ilsley (Berks), Tan (St. Anne's) Hill (Wilts), and other places, still to a large extent adopt the ancient British trackways over the Downs, keeping well above the heads of the coombes or valleys, as did the Welsh drivers of cattle in my remembrance, covering a distance of some 400 miles along the crest of the hills, to avoid the tolls on the high road before the turnpike gates were done away with. Many of these trackways meandering over the broad Sussex and Hampshire Downs were used by the smugglers for carrying their contraband goods from the south coast into the interior of the country, and their course can readily be traced."

Mr. Money, who is familiar with every part of the Berkshire, Wiltshire, and Hampshire Downs, adds: "I think it most probable that the innumerable earthwork enclosures of various forms (where not thrown up as military defences) which stud these open tracts of the Downs, were places wherein to fold sheep and other cattle in safety, and to prevent their wandering by night far away from home, or into the territory of a hostile or predatory tribe. With no other material at hand, an earthwork would naturally suggest itself to the British herdsmen and shepherd."—[*Author's Note.*]

remembered, did accomplish the feat, it is quite a stiff scramble to its summit, from whence you may look down upon Avebury, but a mile away. This colossal effort of prehistoric ages has no equal in Europe ; its purpose still defies the wit of man, and controversy has raged around it for generations. Whether it is the tomb of some matchless chieftain, a monument to some potent god, or merely a part of the great design and scheme of Avebury, will never be revealed. The peasant decided long ago that it was the tomb of King Sel, a monarch with whom I claim no sort of acquaintance.

When I was last at Silbury a shepherd had just accomplished the feat of getting his flock of down sheep through the gate, and as they raced along in a big bunch after the manner of their kind, with bells tinkling, and tearing eagerly at the fresh sample of pasture that grew in the moat,[1] he was taking a well-earned leisure on the bank, his coat and crook and dog beside him, and in sociable mood. "Yes, zur, it's a turr'ble big mound, vor zartin. A nashun sight of volks come yer to look at'n ; I doan't take much notice of her myself, but I've heerd my feyther tell as they druv a hole into her innards onc't, an vound zummat or awther."

And so they did, more than once. In 1723 some workmen planting trees—of which there is now no trace —on the summit, found a skeleton near the surface, which Dr. Stukeley, whose imagination was stronger than his balance in these matters, hailed at once as the great King Sel, and the faith of the shepherd in his majesty waxed more robust than ever. Primary burials, however, lay at the bottom, not at the top of the *tumuli*, and the learned doctor's theories fell to the ground before the cold light of reason.

[1] A writer in *Tit Bits* says that "sheep nod their heads when they are feeding, because they have no incisors or cutting teeth in the upper jaw. The grass is collected and rolled together by means of their long tongue ; it is then firmly held between the lower cutting teeth and a callous pad above, and finally by a sudden nodding motion of the head the little roll of herbage is partly torn and cut off."
Barnes (1569-1607) writes of—
"The sheep, little-kneed, with a quick-dipping nod."
[*Author's Note.*]

BERKSHIRE

SHEPHERDING ON THE BERKSHIRE DOWNS

By J. E. Vincent, 1906

Substantially speaking, the Berkshire Downs are not grazed at all, although it may be conjectured that, if they were, the mutton would be passing excellent. Our Berkshire sheep are not of the hill-climbing type, and they are hardly ever allowed to graze at large. Except along the roads, which they spoil abominably with their cloven hoofs when they are driven from place to place, they take practically no exercise. The system is to pack them on good feed—grass, clover, turnips, or what you will—practically as close as possible, until they have eaten it bare, and then to pass them on to the next plot. Some years ago, for example, I had a rank aftermath of about an acre and a half, which as a favour to me a farmer neighbour permitted his sheep to eat. He hurdled the little area off into two equal plots, put 300 sheep into one of them for one day and night, and to the next for the same time, and behold, every vestige of the grass had vanished. Had they all lain down simultaneously, no ground to speak of would have been visible. That is the way we feed sheep hereabouts, and one rarely sees the Downs dotted with sheep singly or in groups. The Berkshire sheep lives between hurdles, and seldom knows freedom from the day when he or she is born in the sheltered lambing-yard, to that on which the Saxon sheep is converted into the French mutton. Hence comes it that the coat of the Downs is never, or very rarely, that hard, close, and velvet-like turf which one finds where sheep have grazed at their will.

A SHEEP FAIR AT EAST ILSLEY

By L. Salmon, 1909

East Ilsley is a queer little village. At first sight it really appears to consist almost entirely of public-houses.

Shepherd and Flock

The sheep fair of East Ilsley is one of the most important and one of the most ancient in England. The charter for its establishment was granted as long ago as the reign of Henry III., and as the public-houses would not exist were it not for the sheep fairs, so the sheep fairs could not possibly get on without the public-houses. Over the Downs and far away the shepherds have to travel and drive their sheep, and as the fair begins early in the morning, most of the sheep must be there the night before, and the shepherds and the drovers have to sleep somewhere. So they flock within the hospitable doors of " The Star," " The Lamb," and " The White Horse," and thus the necessity of the public-houses is accounted for.

The chief fair takes place annually upon August 1st. You must be at East Ilsley by nine o'clock in the morning if you wish to see it in full swing. All is bustle and life. More than 20,000 sheep are enclosed in pens upon each side of the street, stretching out, a woolly mass, towards the fields behind. Their number lately has been reduced by about one-half from what it formerly was. Thousands of sheep are bleating ; hundreds of dogs are barking loudly, some of them held by the shepherds who are watching over the sheep—others are tied to the pens whilst their owners go a little way off to gossip with their friends, or to make a voyage of discovery through the fair. The dogs are chiefly of the collie breed ; here and there is conspicuous a great English sheep-dog, quieter and more dignified than the somewhat fussy and somewhat mongrel collie.

The passages in between the pens are filled with men in every variety of costume, some of the most picturesque description. Many of the drovers, with their knotted red handkerchiefs and general get-up, remind one of the London coster. Here you may see a man, one of the dealers, in a velveteen jacket and a soft felt hat, looking as though he were dressed for the stage ; there is another in a long white coat reaching to his heels, with huge gold spectacles and white hat, a cross between a bowler and a tall hat cut down. Infinite variety! But the smock-frock—one longs to see it—is, alas, absent. Nevertheless, the endless and picturesque modification which the English costume, prosaic and ugly though one is apt to consider it, seems capable of at a Berkshire sheep fair, is surprising.

One old shepherd, who carries the invariable heavy walking-stick that appears to have taken the place of the shepherd's crook, wears a dark blue smocked coat and a slouch hat. His face bears the calm and beautiful expression that is seen sometimes upon the faces of those old shepherds whose whole life has been spent with Nature alone, in the fields or upon the Downs, with no companionship but that of their sheep and their dogs, when they would wile away the time of their solitude sometimes in carving wonderful designs upon walking-sticks. Now and then in an old house you may see one of these sticks, and an interesting piece of workmanship it is. This man's whole bearing is a combination of quiet dignity and childlike simplicity. When a "snapshot" photograph was taken of him and his great shaggy bob-tailed dog, he came up immediately afterwards, expecting to see his likeness; his face clouded over with disappointment when he was told that a great deal had yet to be done before the picture came into existence. Alas! that picture never did come into existence; one of the manifold misfortunes attending the fate of these precarious things overtook it, and—the opportunity is gone for ever.

Presently an unusually loud bleating of sheep and barking of dogs is heard. What can be happening? The fair is nearly over, and a flock of sheep is being driven, or rather is refusing to be driven away. Sheep are said to have an instinctive and curious objection to going downhill on an unknown road. These particular animals are exhibiting this trait to an amazing extent. A rope has been fastened round the neck of one of them and the victim is being dragged along, in the hope that, sheep-like, the others will follow.[1] Nothing of the kind.

[1] An extreme instance of this tendency in sheep was quaintly exemplified at Kingston-on-Thames one market day. A flock of sheep were being driven through Thames

The hope is altogether a vain one; their conduct is most unsheep-like. They rush and bound about in every direction; they even leap the hurdles and get mixed up with the others. About twenty dogs are now yapping at their heels, evidently thinking it all fine fun. One dog seizes a shepherd by the leg, doubtless in his excitement mistaking him for a sheep. The noise is deafening; the dust is sent flying in smothering clouds. Headlong rush the sheep—in every direction but the one in which they are intended to go, scattering the men, who are powerless to stem the torrent of their mad flight. In the end the sheep get their own way, and go down another road, leading, however, by a circuitous route to the one by which they were originally intended to go. The incident seems to demonstrate very forcibly the powerlessness of man, unarmed, to contend against any animal that chooses to assert its strength.

KENT

A SHEPHERD'S POWER OF ABSTRACTION

By Richard Jefferies, 1887

Beside the path, but just off it so as to be no obstruction, an aged man stands watching his sheep. He has stood so long that at last the restless sheep-dog has settled down on the grass. He wears a white smock-frock, and leans heavily on his long staff, which he holds with both hands, propping his chest upon it. His face is set in a frame of white—white hair, short whiskers, short white beard. It is much wrinkled with years, but still has a hale and hearty hue.

The sheep are only on their way from one part of the farm to another, perhaps half a mile; but they have already been an hour, and will probably occupy another in getting there. Some are feeding steadily; some are in a

Street, when a sheep, seeing itself reflected in the glass of a shop window, rushed at the reflection, broke the glass, and sprang through, followed by the rest of the flock, about twenty in number.—[*Author's Note.*]

gateway, doing nothing, like their pastor; if they were on the loneliest slope of the downs he and they could not be more unconcerned. Carriages go past and neither the sheep nor the shepherd turn to look. Suddenly there comes a hollow booming sound—a roar, mellowed and subdued by distance, with a peculiar beat upon the ear, as if a wave struck the nerve and rebounded and struck again in an infinitesimal fraction of time—such a sound as can only bellow from the mouth of a cannon. Another and another. The big guns at Woolwich are at work. The shepherd takes no heed—neither he nor his sheep. His ears must acknowledge the sound, but his mind pays no attention. He knows of nothing but his sheep. You may brush by him along the footpath, and it is doubtful if he sees you. But stay and speak about the sheep, and instantly he looks you in the face and answers with interest.[1]

SHEPPEY, "THE ISLE OF SHEEP"

By W. Lambarde, 1576

It would seem by the dedication of the name, that this island was long since greatly esteemed either for the number of the sheep, or for the fineness of the fleece, although ancient foreign writers ascribe not much to any part of all England (and much less to this place) either for the one respect or for the other: but whether the sheep of this realm were in price before the coming of the Saxons or no, they be now (God be thanked therefore) worthy of great an estimation both for the exceeding fineness of the fleece (which passeth all other in Europe at this day, and is to be

---

[1] As an illustration of how the whole interest of some of the old shepherds is centred in their sheep, I quote Mr. David Cowen. "A shepherd told me that he had his greatest pleasure in watching his flock, and his master (one of the chief flock-masters in East Suffolk) said that he failed to understand the man who could not sit on a gate for a couple of hours looking at a flock of Suffolk sheep." Prebendary Fraser of Chichester adds: "When I was a deacon I used to visit the old people's wards in a Hampshire workhouse. There was an ancient shepherd there quite worn out and dazed, but he always roused up when the 23rd Psalm was read to him, or if one spoke to him about Archbishop Trench, who, when rector of Itchen Stoke, used to come and talk to him about his sheep."—[*Author's Note.*]

compared with the ancient delicate wool of Tarentum, or the Golden Fleece of Colchos itself) and for the abundant store of flocks so increasing everywhere, that not only this little isle which we have now in hand, but the whole realm also might rightly be called Sheppey.[1]

THE SHEEP OF ROMNEY MARSH

"One of the most singular districts in the south of England."

From *Excursions in the County of Kent*, 1822

By my christendom
So I were out of prison and kept sheep,
I should be merry as the day is long.
King John, iv. 1.

The Marsh is about ten miles in length from east to west; and in breadth, from north to south, about four. Sheep are much more commonly fed on the Marsh than oxen; the number of sheep, according to Boys, is exceeding that of any district in the kingdom. The sheep take their name from the spot where they feed, and are larger than the Southdown, but by no means so large as those of Lincolnshire. Their wool for length and fineness is much esteemed.

THE "LOOKERS" OF ROMNEY MARSH

By ARTHUR FINN, 1910

Romney Marsh is probably one of the oldest and most thickly populated sheep districts of England. The shepherd here is usually called a *looker*,[2] his duties being

---
[1] Sheep are still numerous here. The Isle of Sheppey and the marshes in this part of Kent are somewhat similar to Romney Marsh. In both districts there is a sad lack of shade and shelter for the flocks.—[*Author's Note.*]

[2] Various authorities define "looker" as follows:—Looker, a person who looks after sheep and cattle in marshes and enclosed lands, peculiar to eastern Sussex; also in Kent.—*Glossary of the Provincialisms in use in the County of Sussex.* By Cooper, 1853.

Looker, a looker or superintendent. Essex.—From *The Reports of the Agricultural Survey*, 1793-1813.

Looker, a bailiff. Essex (Foulness). It is also used generally in Essex. A shepherd. Kent (Romney Marsh).—*Annals of Agriculture*, 1784-1815.

Looker, a herdsman. Sussex.—Wright's *Dictionary of Obsolete and Provincial Sayings.*

Looker, a shepherd or herdsman. South.—HALLIWELL.—[*Author's Note.*]

to look after the sheep all through the year. Originally (even more so than now, when it is customary for some stock-owners to keep their own individual men in charge of a distinct business) the looker was paid 1s. 6d. per

Photograph by Edward Lovett.
BACKSTAYS

"The 'lookers' or shepherds of Romney Marsh wear these flat wooden shoes over their ordinary ones when crossing the shingle which extends for some miles near Dungeness. The local name for this strange footgear is 'backstays.'"

acre a year, and might be employed by several different flock-owners, and this is still the custom here. Before shearing, the sheep are washed by being thrown into what is called a "sheep-tun," that is, a shallow well with a gangway for them to pass out by. The Marsh is intersected by "ditches" filled with water. There is an old saying here that "it requires no cleverness to be a grazier provided a man could keep himself from falling into the Marsh ditches." The town of Lydd in old times had its own common flock. There was a town shepherd appointed, who annually accounted to the Corporation for the profits and losses on the feast of Saint Mary Magdalene. Sheep of the Romney Marsh breed go all over the world, notably to New Zealand, Australia, the Argentine, and Falkland Isles; Newfoundland is now inquiring about them. Hardy enough to stand the bleak climate of the Marsh, they will thrive almost anywhere.[1]

EAST ANGLIA

THE NORFOLK BREED

By The Author

The Norfolks were progenitors of the modern Suffolk sheep. Frederic Shoberl, in *The Beauties of England and Wales*, writes of these sheep in 1813: "The Norfolk, or, as it might with greater propriety be denominated, the Suffolk breed of sheep, since the most celebrated flocks are found about Bury, is diffused over almost every part of the county. This race is deservedly esteemed for the fineness of the wool (which is the third in price in England), for endurance of hard driving, and for being the best of mothers. These excellences are, however, counterbalanced by their voracity, a want of tendency to fatten, resulting

[1] A Kent landowner, writing in 1910, says: "Romney Marsh keeps a lot of stock in the summer months, but the land will not support so many in the winter. The lambs here are born in April, and by September are all *out*—'out to keep'; they are sent either to the uplands of Kent or to Sussex, Surrey, Hampshire, Bucks, etc. They are sometimes driven up and distributed on the way, but the more modern graziers send by train, and the animals are then handed over to the farmers by an agent. This is my plan. I send to another county about 3000 in the autumn. Some send sheep away as well as lambs, and reduce the land in the winter to two sheep to the acre."

One who witnessed this migration of sheep many years ago describes it as a wonderful sight, thousands of sheep being *led* along a country road near Tunbridge Wells, raising clouds of dust; the air smelling of sheep. The migration of sheep is mentioned in the ancient *Berkeley Records*.—[*Author's Note*.]

Photograph by Clarence Hailey & Co., Newmarket.

NORFOLK EWES AND LAMBS

The property of Mr. Dermot M'Calmont of Crocksford, Newmarket. Part of the only flock of Norfolk sheep of any pretension in England, numbering about a hundred.

Photograph by Clarence Hailey & Co., Newmarket.

A RAM OF THE FLOCK

Shepherd and Flock

from an ill-formed carcase, and a restless and unquiet disposition. The Norfolks were eventually superseded by Southdowns, introduced by Arthur Young, who gives the number of sheep in Suffolk as 240,000."

In 1910 the Norfolk is described as "a very highly bred animal, somewhat wild in nature, disliking confinement, particularly artistic in appearance, and the best mothers of any known breed of sheep."

LINCOLNSHIRE

LINCOLNSHIRE LONGWOOLS

By The Author

The county of Lincolnshire enjoys the reputation of having more sheep than any other county of England. In fact, the sheep of Lincolnshire are believed to aggregate more than a million in number. Mr. Rider Haggard remarks in *Rural England* (1902) : " In Lincolnshire sheep are everywhere, on the high land and the low." In some parts of the county, however, notably the southern marshes, fewer are kept every year ; the farmers devote all their energies to potatoes and mustard. The Lincoln longwool is pre-eminent as regards breed, single specimens having been sold at auction for more than a thousand pounds.

THE MIDLANDS

"LEICESTERS" AND OTHERS

From *Pictorial Half-Hours*, 1850

In Leicestershire, Nottinghamshire, Warwickshire, Derbyshire, and throughout the Midland counties generally, the old long-woolled sheep, essentially the same in character, were large, gaunt, clumsy animals. . . . The farmers were content—nay, more, they scorned all ideas of improvement as visionary. Science was not the handmaid of practical farming, the land was not economised, and time, money, and food were wasted upon animals which ill repaid the farmer for his trouble and outlay. But the dawn of a revolution was at hand. Chemistry, geology, and physiology had reared up temples in the land, whence radiated a light before which the clouds and mists of ignorance began to be dispelled. The genius of agriculture awoke from a long trance, and a new era commenced. Old plans became obsolete, and as a new system developed itself, so in a parallel ratio were improvements instituted in the breeds of horned cattle and of sheep over the country generally. . . . From that time the star of the old unimproved races began to set.[1]

YORKSHIRE

HENRY CLIFFORD, "THE SHEPHERD LORD" OF LONDESBOROUGH

From *The Book of Days*. Edited by ROBERT CHAMBERS, 1869

The life of Henry Clifford, commonly called the Shepherd Lord, is a striking illustration of the casualties which attended the long and disastrous contest between the Houses of York and Lancaster. The De Cliffords were zealous and powerful adherents of the Lancastrian interest. In this cause Henry's grandfather had fallen at the battle of Towton, that bloody engagement at which nearly 40,000 Englishmen perished by the hands of their fellow-countrymen. But scarcely had the Yorkists gained this victory, which placed their leader on the throne as Edward IV., than search was made for the sons of the fallen Lord Clifford. These were the two boys, of whom Henry, the eldest, was only seven years old. But

[1] The Leicester breed of sheep first came into notice in England in 1755. Robert Bakewell (1726-1795), the agriculturist who introduced the famous Dishley or new Leicester sheep, had a quaint arrangement when he showed his sheep on his farm, which T. Britton thus describes in *The Beauties of England and Wales*, 1807 :—" They were exhibited singly in a small house adapted to that purpose, having two opposite doors, one for admission, and the other for retreat ; and the inferior were always introduced first, that the imagination of the inspector might be raised by degrees to the utmost pitch at the exhibition of the last and finest."—[*Author's Note.*]

Shepherd and Flock 57

the very name of Clifford was so hated and dreaded by the Yorkists, that Edward, though acknowledged king, could be satisfied with nothing less than the lives of these two boys. The young Cliffords were immediately searched for, but their mother's anxiety had been too prompt for the eagerness of revenge. They could nowhere be found. Their mother was closely and peremptorily examined about them. She said she had given direction to convey them beyond the sea, to be bred up there ; and that being thither sent she was ignorant whether they were living or not. This was all that could be elicited from their cautious mother. Certain it is that Richard, her younger son, was taken to the Netherlands, where he shortly afterwards died. But Henry, the elder, the heir to his father's titles and estates, was either never taken out of England, or, if he were, he speedily returned, and was placed by his mother at Londesborough, in Yorkshire, with a trustworthy shepherd, the husband of a young woman who had been under-nurse to the boy whom she was now to adopt as her foster-son. Here in the lowly hut of this humble shepherd was the heir of the lordly Cliffords doomed to dwell—to be clothed, fed, and employed as the shepherd's own son.

In this condition he lived month after month, and year after year, in such perfect disguise that it was not till he had attained the fifteenth year of his age that a rumour reached the court of his being still alive in England. Happily the Lady Clifford had a friend at Court, who forewarned her that the king had received an intimation of her son's place of concealment. With the assistance of her then husband, Sir Lancelot Threlkeld, she instantly removed "the honest shepherd with his wife and family into Cumberland," where he took a farm near the Scottish Borders. Here, though his mother occasionally held private communications with him, the young Lord Clifford passed fifteen years more, disguised and occupied as a common shepherd ; and had the mortification of seeing his castle and barony of Shipton in the hands of his adversary, Sir William Stanley, and his barony of

Westmorland possessed by the Duke of Gloucester, the king's brother.

On the restoration of the Lancastrian line by the accession of Henry VII., Henry Clifford, now thirty-one years old, was summoned to the House of Lords and restored to his father's titles and estates. But such had been his humble training that he could neither read nor write. The only book open to him during his shepherd's life was the book of Nature; and this, either by his foster-father's instruction, or by his own innate intelligence, he had studied with diligence and effect. He had gained a practical knowledge of the heavenly bodies, and a deep-rooted love for Nature's grand and beautiful scenery. Having regained his property and position, he immediately began to repair his castles and improve his education. He quickly learnt to read and write his own name; and, to facilitate his studies, built Barden Tower, near Bolton Priory, that he might place himself under the tuition of some learned monks there, and apply himself to astronomy, and other favourite sciences of the period.

Thus this strong-minded man, who up to the age of thirty had received no education, became by his own determination far more learned than noblemen of his day usually were, and appears to have left behind him scientific works of his own composition.

His training as a warrior had been equally defective. Instead of being practised from boyhood to the use of arms and the feats of chivalry as was common with the youth of his own station, he had been trained to handle the shepherd's crook, and tend and fold and shear his sheep; yet scarcely had he emerged from his obscurity and quiet pastoral life when we find him become a brave and skilful soldier—an able and victorious commander. At the battle of Flodden he was one of the principal leaders, and brought to the field a numerous retinue. He died on the 23rd of April 1523, being then about seventy years old.

SHEPHERDS' MEETING-TIME

From *The Globe*, 1878

On November 7th the autumn gathering of the moorland shepherds of the north of England took place in the village of Saltersbrook, Yorkshire, and as an indication of the value of such meetings to farmers, it may be stated that no fewer than 121 strayed sheep were returned to their rightful owners. In the spring the farmers turn out immense flocks, sometimes numbering thousands, to nip the tender but scant herbage which the moors furnish. These sheep will wander about the whole of the summer, and it is the shepherds' duty to see that they do not stray. Occasionally animals will wander miles from the parent flock, to be picked up and kept by other shepherds until "meeting time" comes round, when they are enabled by the marks stamped on their coats to return them. The shepherds have not such an easy time of it as many would imagine. They have frequently to walk very long distances. Sheep have an awkward habit of getting upon their backs, and when once in this position they must lie there and die unless somebody comes and turns them on to allfours.[1] Then again whole flocks of sheep will find their way into some deep ravine, out of which they are too simple to pick their path. There, in the absence of grass (for in many of these gullies nothing is to be found but bracken and huge moss-covered boulders), they will stand bleating until the shepherd opportunely arrives and drives them out. The wages the men receive are very small, averaging in Yorkshire about 17s. a week. In the summer many of these shepherds sleep out night after night, preferring, when tired and a long distance from their homes, to throw themselves down to rest among the heather. It is from these shepherds that in the spring and autumn months reports come as to the prospects of grouse shoot-

---

[1] A sheep is said to be cast when it lies on its back. In some parts the term is "awelted," in others "rigwelted," *i.e.* turned over as a "ridge" of earth by the ploughshare. In Orkney the word *aval* is used. O.N. *af-velta*.—[*Author's Note.*]

ing.[1] " Meeting time " is the shepherd's carnival. Quiet and steady enough generally, he then gives rein to his bacchanalian proclivities, and under the mask of merriment makes himself thoroughly miserable till he gets back to the moors again.

THE LAKE COUNTRY

SHEEP-FARMING IN CUMBERLAND

By J. BRITTON AND E. W. BRAYLEY, 1802

Sheep-farming, which is prevalent in the mountainous tracts and on the borders of large commons, seems to be less understood here than in other parts of the kingdom. To account for this is easy, and the reply of a simple husbandman is sufficient for the purpose. "At Penruddock," say the persons who drew up the agricultural report, "we observed some singularly rough-legged, ill-formed sheep; and on asking an old farmer whence the breed was obtained, he replied: 'Lor', sir, they are sik as God set upon the land; we never change any.'" These sheep-breeders are generally so attached to their own kind that they seldom care to make those experiments which the perfection of the science renders necessary.

THE HERDWICK

By A. G. BRADLEY, 1901

From Skiddaw to Black Combe, from Ennerdale to Shap, the Herdwick sheep is the pivot on which all local life, not wholly absorbed in the tourist business, turns. . . . Nothing, strange to say, is more like the grey colouring of the fleeces grown on these pure sweet fells than that you sometimes see on sheep pastured amid the smoke of a great city, and it is remarkable that the wool of the Herdwicks

[1] The Highland shepherds say that the little sheep-tracks are invaluable for the grouse, as the young birds can run on them and not be strangled by the long grasses.—[*Author's Note.*]

bred in Wales turns to pure white again. The lambs are mostly piebald, black and white, with a humour of appearance all their own, though their fleeces tone down afterwards to the right shade. Farmers tell me, however, that the tendency to black wool is always very strong with Herdwicks. They are apt, moreover, to run to horns in the ewes, which (from a breeding point of view) is incorrect.

A Curious Usage in the Lake District
1901

The Gatesgarth Fells are no longer "stinted pasture" or common land. The landlord has fenced the boundary, a not uncommon proceeding nowadays, and of infinite convenience to the farmer. For, though the mountain sheep on commons like the Helvellyn and Matterdale ranges acquire an hereditary instinct for keeping to their own beats that is wholly marvellous, a certain amount of straying is inevitable. This very custom of a common grazing-ground, to which each valley farm has a recognised but unrecorded right to send up so many sheep, has maintained, and indeed necessitated, a usage that is quite unique in England—namely, that of the flock being the landlord's property; a fixture, in fact, of the farm, and allowed for in the rent paid, if not, indeed, the principal factor. The origin of this is that only sheep bred upon the mountain know the range. A strange flock would give such trouble for so long a time that an incoming tenant would be almost compelled to purchase an outgoing one. Hence the sheep, by a natural process, become part of a property that without them is useless. They are valued to the tenant and periodically appraised, the difference at his death or outgoing being due from the one party to the other. The farmer therefore, under this system, is in literal fact a shepherd working on the profit system with his landlord, though the annual payment in the shape of a rent is a fixed one. Gatesgarth, though now fenced in, is held upon these lines, the father of the present occupant having been a noted breeder and prizewinner.

ANOTHER ACCOUNT OF THE HERDWICK

From The Morning Post, 1909

If we were asked to name the hardiest race of sheep kept in this country, we should unhesitatingly give preference to the Herdwick. This breed, a small-horned type of longwool, lives amongst the crags of the beautiful Fell district of Cumberland. It differs from every other type of sheep reared in England, taking many years to mature, and picking up a living where even the hill blackface race of Scotland would starve. History credits it with descent from sheep that escaped from a wrecked Spanish galleon when Drake scattered the proud Armada; but doubt may be thrown upon that supposition, which has been conjecturally used by chroniclers whose researches have led to no definite conclusion. The same theory is held with regard to the blackface sheep of Scotland.[1] . . .

Although in its later years the Herdwick is a light-coloured sheep, the lambs are born with black faces and legs, the only bit of white apparent being round the tips of the ears. At three years old the sheep has changed from a dark colour to a steel-grey. The rams may or may not be hornless. They are usually horned, but the ewes are polled. . . . The flocks are wintered on the lowlands, but their home is on the bare rock-crowned hills.

ISLE OF MAN

THE TWO BREEDS OF THE ISLAND SHEEP

By John Feltham, 1798

The native stock is small and hardy, and would endure the roughest weather with little loss, and the meat tasted fine. This is still the mountain breed. There is also a peculiar breed, called laughton,[2] of the colour of Spanish snuff, and these are not so hardy, and more difficult to

---

[1] The Manx folk have the same tradition respecting their loaghtan, or "laughton," sheep.—[*Author's Note.*]
[2] For spelling see page 64.

fatten. The natives like the cloth and stockings made of the wool.

THE "LAUGHTONS" OR "LOAGHTANS"

From *The Sheep*, 1837

The sheep are small on the hills, seldom exceeding eight to ten pounds the quarter, and producing fleeces of short or middle wool weighing two and a half pounds. They have much resemblance to the Welsh sheep, and have most of their peculiarities and bad points. They are narrow-chested and narrow-backed, long in the leg, and deficient in shoulder. They are found both horned and polled, mostly of a white colour; but some of them are grey, and others of a peculiar snuff or brown colour, termed in the island "laughton" colour. This colour, either covering the whole of the sheep or appearing in the form of a patch on the neck, is considered as the peculiar badge of the Isle of Man sheep. In the valleys a larger sheep with longer wool, a proper long-woolled sheep, is found. The flesh of both breeds is said to be good, and the wool of the hill sheep valued in the manufacture of stockings and some of the worsted goods.

SHEEP AND SHEPHERDING IN THE ISLE OF MAN

By THE AUTHOR

Sheep used to keep their own places upon the mountains without fence of any kind, but the land is now enclosed. The Crown has also enclosed the commons, which are let. Before "disforestation of the commons" in 1860, any one, by paying a nominal rent, might send sheep to feed on the mountains. A good many of the shepherds are Scotsmen. The Manx shepherds turn their hand to any farm-work, and the farms and flocks not being large several farmers act as their own shepherds, assisted by a son or some other farm-hand. The sheep when taken from the mountains have a way of returning to their old haunts, as is the custom with sheep in Wales and elsewhere. This is very tiresome when there are other plans for them, or when they have changed owners. The name for one of these places is *oayll* (a haunt, a place frequented). The Manx shepherd often makes a pet of some of his flock, and particular favourites may be seen running to meet him as soon as he appears. On the mountains are small yards made of stone without mortar, and called *paabs*; these are used for catching sheep in, and would seem to be much the same as the " rude enclosures " of Shetland, which are called *punds*.[1] Both in Shetland and in the Isle of Man it was the custom to hunt the sheep with dogs.

There are three kinds of sheep in the island :—1. The loaghtan, or "laughtan," as it is variously spelt (loagh is pronounced like the Scotch loch), which is the name popularly given to the brown flocks of the old Manx breed, though, as elsewhere pointed out, it is really the name of a colour, not of the breed itself. 2. A white sheep with a yellow face.[2] 3. The *keeir* or black sheep, a mouldy grey (*keeir* in Manx means dark grey). The loaghtan, or rather *lughdoan*, which is the correct spelling according to Cregeen's *Manx Dictionary*, is, he tells us, derived from *lugh* (mouse) and *dhoan* (brown), these colours when mixed producing the shade which is understood by loaghtan ; it cannot be from *lhosht dhoan* (burnt brown), though this better describes the colour. The loaghtan is said by some to be a purely Manx breed and only known to the islanders by the Manx name. This, however, is not correct, for similar sheep are to be found in Shetland, though not in Orkney, and a few also in Scotland. When the pretty soft brown wool of the loaghtans becomes weather-beaten it gets lighter at the edges. There sometimes occurs in a loaghtan flock white sheep with patches of brown, but when a flock of loagh-

---

[1] See page 97.
[2] Sacheverell, in *Account of the Isle of Man*, published 1707, writes of some sheep on the island " of a yellow or rather buff colour."

Shepherd and Flock 65

tans is named it is understood that brown sheep are meant. They have two and in some cases four horns. The race had become almost extinct in the Isle of Man in the earlier part of the last century, their place being taken by larger sheep brought over by Scotsmen who rented Manx commons. A Manxman in Baldwin, Quirk by name, who was of a conservative disposition, kept some of the old stock. He was quite a character, and believed, as he put it, "the oul' times were bes' for all." He ploughed

By W. Marsden.   From a painting in possession of Mr. J. C. Bacon.
LOAGHTAN EWE AND RAM

with oxen, and would not have any new-fangled English improvements about his farm, but held on with his poor little loaghtans while his neighbours went in for the larger English and Scotch breeds which paid them better. Thus the loaghtans were being gradually displaced for such breeds as the Shropshire and Leicester in the lowlands, and by the Scotch mountain breed in the highlands. Many years after, Colonel Anderson, who still lives in the island, on passing through Baldwin, high up in the fastnesses of the Manx mountains, there saw these sheep. He brought some to his place at Michael, and later on other landowners also bought some of them. In these days when picturesque, even if less useful, types are gradually passing away, it is refreshing to hear of conservative Quirk and those who followed his example. Mr. J. C. Bacon, of Santon, Isle of Man, writes : " I have a brown flock of pure Manx mountain sheep. The word loaghtan only refers to the brown colour, which is what the Shetlanders call moorit or moor-coloured. It is, however, not really the name of the breed but of the colour. Our Manx sheep are very small and finely shaped, a well-defined and handsome variety of a breed which for want of a better name I call the short-tailed sheep of Northern Europe. They were no doubt at one time wild ; and on the island of Soa, one of the St. Kilda group, are still practically in a state of nature. The domestic short-tailed sheep is found pure in Iceland, the Faroe Isles, the Shetlands, Isle of Man, and a few in the Outer Hebrides, where I saw one or two good specimens this year. Probably there may still be some in the more remote Scotch highlands and in Ireland. The breed runs in peculiar colours—white, black, brown, grey, and frequently in piebald mixtures of these colours. They have a tendency to produce four horns, and sometimes have even five or six. They are very hardy and picturesque, make excellent mutton, and except in very cold climates yield wool of exceptional quality. The Shetland shawl-wool is, of course, renowned. One peculiar feature of this breed is that the tail never reaches the hocks, and another is, that if not shorn they cast their fleece in summer. The Shetlanders, Faroes, and Icelanders never clip these sheep, but simply pull the wool off when it becomes loose. Black four-horned sheep found in some English parks are invariably described as St. Kildas, and some no doubt came originally from there. It is said, however, that the only sheep there now are the Scottish blackface, and the little brown, nearly wild flock just mentioned as occurring on the Isle of Soa."

OLD ENGLISH BREEDS OF SHEEP

By Walter Skeat, 1910

There seems to be little doubt that the Manx mountain breed represents one of the most ancient of the old English breeds of sheep. Remnants of a similar breed occur in some parts of Scotland, the Shetlands, and Hebrides, and sheep of a black colour were also very common in ancient Ireland. Giraldus Cambrensis (A.D. 1200)

MR. J. C. BACON AND HIS MANX "LOAGHTAN" SHEEP

states that the dress of the Irish at that time was "generally black, the sheep of the island *being of that colour.*" In other words, the common people wore garments made of the natural black wool. And we know further that the ancient Irish sheep, generally at least, were of the many-horned variety. It is practically certain that they belonged to the "black" many-horned breed here discussed, or at least to some similar breed. In the illuminations of Anglo-Saxon manuscripts the horned variety are oftener represented than the polled sheep. Altogether it seems fair to assume that this dark breed anciently existed in England as well as in many other parts of the British

Isles, where it survives, or has survived till recently. I was speaking one day to Mr. E. Magnusson (the well-known Icelandic scholar), and mentioned the fact that a many-horned sheep was formerly found in Ireland. Mr. Magnusson replied : " That is a fact of great interest to us Icelanders, because many-horned sheep sometimes of a black colour are still to be found in Iceland, and I did not know that they were to be found elsewhere." And Mrs. Magnusson kindly hereupon showed me some

ST. KILDA SHEEP. THE PROPERTY OF MR. D. MACPHERSON.

"Many of the St. Kilda sheep are black and four-horned. The wool, which in its rough state is called 'clip,' is loaded with fat. After scouring, 24 lbs. will only weigh 14 lbs., so much weight is lost."

black Icelandic wool which she had obtained for spinning, an art which she still practises. Again, " There is no doubt," Dr. Chalmers Mitchell remarks, " that a blackfaced sheep generally with dark brown wool was formerly common in many parts of Great Britain, and these sheep had usually four horns." The St. Kilda or Hebridean four-horned sheep are the best-known examples, but others turn up from time to time from different parts of the country.

Mr. C. I. Elton, in his *Origins of English History*

(1890),[1] has the following important passage :—"The Celts in the Midland districts [long after Caesar's time] may have lived in permanent villages, raising crops of oats or some rougher kinds of grain for food, and weaving for themselves garments of hair from their puny, many-horned sheep. But the ruder tribes who subsisted entirely by their cattle would naturally follow the herd, living through the summer in booths on the higher pasture-grounds, and only returning to the valleys to find shelter from the winter storms. Professor Rolleston states that no one with the evidence before him 'can doubt that the goat, sheep, horse, and dog were imported as domesticated animals into this country in the earliest Neolithic times.'[2] It has been questioned whether the sheep was known in these islands before the Roman invasion, chiefly because it is difficult to distinguish its remains from those of the goat. But the latest discoveries are in favour of the theory that the goat had been to a great extent superseded by the sheep as early as the beginning of the British Age of Bronze."

WALES

SHEPHERDING IN ANCIENT WALES

From *Ancient Welsh Husbandry*
(A Commercial and Agricultural Magazine)

Sheep ought to be housed in the beginning of spring when they are bringing forth lambs, and in winter they should be turned to places under the influence of the sun ; and thou art not to fold them too much on fallow land. Shear them at Michaelmas, so that the marks of the scars may disappear upon them against the winter, and do not milk them later than August.

HABITS OF WELSH SHEEP

By THEOPHILUS JONES, 1809

Here our sheep resemble their aboriginal masters in their manners and mode of life. While they are de-

---

[1] Quoted by permission of Mr. Bernard Quaritch and author.
[2] *British Barrows*, p. 750.

pastured in fields and lowlands, and have boundaries prescribed to them, they have a mischievous activity which baffles human ingenuity to correct. Place them on a mountain where they are apparently free and may roam whither they please, and they stick to a favourite spot as if they were surrounded by a wall. After they have been accustomed to graze upon a particular part of a mountain, if they are not disturbed when at rest at nights, they are prisoners by choice, and cannot be removed from thence without difficulty. This is perfectly well understood by the proprietors of sheep in this country, who sometimes avail themselves of their knowledge in a very artful manner. When there is a right of intercommoning, which is frequently the case here, the shepherd who wishes to prevent a new flock from depasturing on the same bank or hill with those called the *old settlers*, comes at the dusk or in the middle of the night, rattles some stones which he carries in his pocket, throws up his hat, or takes up clods and throws them about him in all directions. This, one would suppose, disturbs his own sheep as well as his neighbour's. It is, indeed, particularly disagreeable and unpleasant to both; but the new settlers not being so much accustomed, and of course not so attached to the spot, give up the walk, and leave it in the sole possession of the old occupiers.

There are also some other traits in their character deserving of notice. When they are first driven to the hills from the low grounds the old sheep, with that affection which is, however, not peculiar to this animal, mount to the highest eminence, and leave, or rather confine, the yearlings and youngest to the lowest part of the hill, showing them by their conduct, perhaps informing them in their language, that they are not so capable of enduring cold as those who have been accustomed to a more bleak and elevated situation. It is very certain also that Providence has implanted in them for the preservation of their species a *presentiment* of the approach of hard weather, particularly of snow, sometimes so fatal to them. A day or two before it falls they are observed to avoid the ditches and other situations where drifts are likely to be found, and sometimes, though seldom, they have been known to quit the hills entirely, to overleap the enclosures, and to come down into the vales a day before a storm commenced.[1]

There is also a peculiarity (as it is said) in the sheep bred in Glamorganshire, when sold and delivered into Brecknockshire, which is very remarkable; but, incredible as it appears, it is attested by the universal voice of those who are conversant with this species of traffic. They assert positively that if a lot of sheep be brought from the former county into the latter, the purchaser is obliged to watch them for a considerable time more narrowly, and with greater care, than any other part of his flocks. They say that when the wind is from the south they *smell it*; and, as if recognising their native air, they instantly meditate an escape. It is certain, whatever may be the cause, that they may be descried sometimes standing upon the highest eminence, turning up their noses and apparently snuffing up the gale. Here they remain, as it were, ruminating for some time; and then, if no impediment occurs, they scour with impetuosity along the waste, and never stop until they have reached their former homes.

SHEEP CHARACTER IN CARNARVONSHIRE

By the Rev. J. EVANS, 1812

The sheep are the ancient Alpine sort, unadulterated or unimproved by any foreign mixture, and form a distinct and very curious breed. They are compared with the Cotswold, Leicester, or even Ryeland breeds, very diminutive animals, and far inferior in size to those of the adjacent county of Anglesey. Some of them in symmetry resemble the *Merino* breed of Spanish sheep. Like the latter, they are migratory, though not to an equal extent, travelling up to the mountains during the summer months, and at the approach of winter descending to the lowland pastures. . . .

[1] *Infra*, "Folk-Lore," p. 315.

From their ranging mode of life these sheep assume a very different character and habitude to those of an enclosed country. They roam wherever inclination leads, confined by no fences; and, frequently unattended by a shepherd, are obliged to have recourse to their own exertions against their formidable enemies the foxes, who here assemble in troops, and the ravens and large birds of prey, who from necessity in this grainless country become carnivorous. The sheep themselves appear quite different animals. Instead of assembling in large flocks they form gregarious parties, generally consisting of ten or twelve, of which number, while feeding, one stands at a distance as sentinel to give notice of approaching danger. If the guard perceive anything making towards the little flock, he turns and faces the enemy, and permits him to advance within about one hundred yards. If his appearance be hostile and he continues to advance, the guard then warns the party by a shrill whistling noise till they have all taken the alarm, when he joins them in the rear, and they all betake themselves to the more inaccessible parts of the mountains.

SHEEP OF THE SNOWDONIAN RANGE

By The Author

The Rev. Morris Griffith, writing from Anglesey of Welsh sheep, says: "It is a well-known fact that sheep when removed from their accustomed feeding ground will make an effort to get back to their haunts. Thousands of sheep belonging to different farmers feed on the Snowdonian Range, and in June they are brought down to be shorn; after that operation they are taken back, each flock going straight to its accustomed place and remaining there. Even though in their progress they have to pass through other flocks, they never mix." Mr. W. B. Gardner corroborates this, adding: "Sheep bred on a farm and kept till they are, say, five years old, no matter where they are taken, never forget their old home; this is true of all breeds. Drive them to the market, then take them a distance of eighty or a hundred miles away from their 'native heath,' and the morning after arrival at their new pasture they will be found huddled together at the points nearest the hills and straths they have left, and wearing looks of deep dejection. But for artificial and natural barriers they would return in unbroken order to the scenes of their youthful days. Also companies or 'cuts' of sheep will return after shearing or dipping to their old ground, and if they have to pass through other flocks the process is similar to that witnessed at a junction where people pursuing various journeys meet."

IRELAND

SHEPHERD AND FLOCK IN ERIN

By Ralph Fleesh, 1909

Sheep-farming, as a successful speculation in Ireland, is largely dependent, as elsewhere, on the skill and devotion of the shepherds, while they in turn rest the competency of their craft upon the celerity and sagacity of the collie. Here Irish flock-masters are extremely fortunate, for the shepherds and sheep-dogs of Ireland are admittedly a credit to the calling. In Ireland, too, the shepherd (who not infrequently is of Scotch extraction) makes his master's interests his own. In a real sense, he lives for his sheep, and, instead of taking orders from his master, gives, under the cover of suggestion, all instructions as to the home management and classing of the sheep for the different markets. A keen and constant student of his flock, he knows the family history of each of its members, and this knowledge is of inestimable value both as regards treatment for disease and for pedigree purposes. The Irish shepherd, like his Scotch and English brethren, seldom indicates a desire for change, unless his bank-book gives ground for the ambition of becoming a master himself.

Even then he hesitates, for the pleasure he derives from the discharge of his present duties, which are all more or less masterful, is unmarred by financial worries, and makes him wonderfully contented. If any new departure in his life be made, it will certainly be in the direction of sheep-farming, since to cast aside the crook, say, for the plough, or for a lordship over a few cows, would mean a serious step down, amounting almost to disgrace. Proud of his calling, the supreme desire of the Irish shepherd is to live till the end comes with his dogs and sheep.

The "pack" system of payment (*i.e.* the ownership of a number of sheep which graze on the farm) has almost disappeared. Now the farmer simply pays the shepherd a wage—one cow, sometimes two, being allowed. In the west of Ireland the remuneration is unjustifiably low; but in other parts I know of shepherds who are in receipt of £75 a year, plus a cow and a "follower," or calf. This not only ensures a fair measure of domestic comfort, but renders it possible for the shepherd to make some small provision for the needs of old age.

Far removed in most cases from the busy populous centres—indeed, a near neighbour is the exception rather than the rule—the shepherd, his wife, and family, with a resource that is truly wonderful, form a little community of themselves, and keep alive parochial and civic interests in a way that neither their situation nor numbers would lead a casual observer to expect. The result is achieved in this way: dogs, cows, "pet" sheep, hens, even the swine, all have their appointed spheres in the little pastoral kingdom among the hills. Thoreau's *Walden*—one of the books of Nature's Bible—is but the glorified picture of a shepherd's home.

The collie, as is his right, receives more attention than any other of the company of subjects. For it cannot be denied that he is the chief breadwinner, and that of all the Irish shepherd's belongings he is the most indispensable.

Although Ireland is strongly assertive in her claims of nationality—and these claims we are not to examine, far less dispute—there is now nothing distinctly national about

Shepherd and Flock

the breeds of sheep seen within her borders.[1] Indeed, the impression left after wandering over the face of pastoral Ireland is that sheep-farming, as we rate the industry to-day, is a comparatively modern development—if development it can be called—of agricultural enterprise in the country, and that originality in breed or system of management is consequently not to be expected. With a few exceptions, all seemed to be at an experimental stage, although, doubtless, many native agriculturists, sharing the proverbial patriotic fire of the Celt, would keenly contest this finding. Still, it has been given a very subordinate place in the sheep-farming world. That there is a social and consequently an economic reason for this, all students of Irish history know. But the growth of industry and corresponding increase of population (if not in Ireland, then elsewhere not far distant) stimulated Irish farmers to fresh and greater effort in the production of mutton and wool, a similar impetus having been given to sheep-farmers, and those with money invested in general agriculture, in all parts of Great Britain.

The blackface sheep, whose origin is somewhat problematical, but whose favourite pastures for many ages have been the hills and dales of Scotland, has crept into the good graces of Irish agriculturists, and promises to become, if it has not already done so, the chief of the fleecy quadrupeds. Many explanations of this rather phenomenal growth have been offered, but the most reliable, since it is the one which commands the respect of economic science, is that the preference for blackface mutton, particularly in Scotland, was so pronounced that for a long time other breeds were comparatively neglected. This preference, though not so marked to-day as it was, has left a history and created a taste, which factors have been mainly instrumental in giving sheep a place in the artistic world. This charmed realm of "infinitudes" is now the chosen sphere of blackface sheep-breeders. Border Leicesters are also found in Ireland. Splendid types may be seen on a number of farms. Then there are Lincoln longwools, Shropshires,

[1] See page 76, "Ancient Irish Four-Horned Breed."

Hampshire Downs, and the Roscommon sheep, an animal that suits certain districts better than any other breed. As a rule flock-masters of Ireland ship their wool to Glasgow. The manufacture of Irish tweeds, however, has become a considerable industry, several experts having given it as their opinion that the quality of goods produced would ensure a much larger market were the machinery improved and certain restrictions removed. Irishmen when speaking in praise of their home-made tweed are wont to aver that it is "unwearable," meaning, of course, that it never wears out.

SHEEP IN ANCIENT IRELAND

By The Author

The ancient Irish four-horned breed mentioned by P. W. Joyce in his *Social History of Ancient Ireland*[1] (vol. ii. p. 280) as having long been extinct, was probably for the most part identical with the black breed actually recorded by Giraldus Cambrensis, though white sheep were also known. To this Dr. Joyce adds that sheep were kept everywhere, as they were of the utmost importance, partly as food and partly for their wool; and they are constantly mentioned in the Brehon Laws as well as in general Irish literature.

Pasturage was in common, the ground so reserved being mountain land as a rule, and unfenced. The arrangement under which in proportion to the size of the farm the right to grazing was regulated, took a cow as the legal limit:

1 cow = 2 two-year-old heifers.
1 two-year-old heifer = 2 one-year-old heifers.
1 one-year-old heifer = 2 sheep.
1 sheep = 2 geese.

The oldest form of the Irish sheep-house, into which the sheep were driven at night, took, like the pig-house,

[1] Messrs. Longmans, Green & Co. (by permission).

the form of a round shed, which was placed within the circular stockade or "rath" of the ancient Irish farmer (Joyce, ii. 41). The milking of ewes was a common practice in ancient Ireland, as has been the case also in other parts of Britain (in England and Scotland) from the earliest times. And the first day of spring, which began on the 1st of February, was actually called "Oi-melc," or Ewe-milking, "in the pagan times, the heathen name giving place to St. Bridget's Feast Day." Lilting or milking tunes were also commonly used. About the year 1430 the *Libel of English Policie* (p. 199) enumerates "Skinnes of Otter, Squirell, and Irish Hare, of Sheepe, Lambs, and Foxe," among the "chaffare" or merchandise of Ireland.

CONNEXION BETWEEN THE IRISH AND FAROES' BREED

Mr. Nelson Annandale, in *The Faroes and Iceland*,[1] tells us that the majority of the first settlers in the Faroes came, not direct from Scandinavia, but from the British Isles, where some of them had been in residence for two generations. "The Suderoe folk often say that they are of Irish, or rather 'Westman,' origin : and the ' men of the West' in old Norse history includes both the inhabitants of Ireland and those of the outer Hebrides. They also gave a name to Westmannhavn on the north-west coast of Stromoe . . . a place which their ships are said to have frequently visited. A certain amount of evidence is given for this view by the fact that a breed of sheep appears to have existed in the Faroes, and especially on the little islands near Suderoe, before the Norse settlement, and, indeed, to have given a name to the group (*fær* = sheep ; *ey* = island).[2] It is impossible that these sheep could have originated in little islands separated by nearly two hundred miles of sea from any other land ; it is unlikely that they are so ancient as any former land connexion which may have existed with this country, or that they could have been introduced by other than human agency, though

---

[1] Quoted by permission of the Delegates of the Clarendon Press.
[2] A more probable sense is *Fair* Island.—[*Author's Note.*]

they conceivably may have been brought by a drifting wreck." The above would account for the evident connexion between the surviving many-horned "black" sheep of the Faroes and Iceland and the similar breeds of Ireland, the Shetlands, and Hebrides, from which they must apparently have come.

In the *History of Quadrupeds* (1781) we find " Boethius mentions a species of sheep in St. Kilda larger than the biggest he-goat, with tail hanging to the ground, horns longer and as thick as those of an ox."[1] (Note that this sheep differs from the four-horned sheep described as short-tailed by Mr. J. C. Bacon.) Also in the same book is named the many-horned breed common in Iceland and other parts of the north, usually with three horns, sometimes with four, or even five.

" In Ireland as an aid to herding, bells were sometimes hung round the necks of cows and sheep. Animals thus furnished are said in the glossary to the *Senchus Mor* to be 'privileged' (Irish, *uaisli*; singular *uasal*, lit. 'noble'), which meant nothing more than that they were distinguished above the rest of the herd. There was a fine for removing the bell. Such bells have continued in use till this day, and in the National Museum may be seen many specimens, some no doubt modern, but some very old."[2] Mr. J. C. Bacon informs me that he has never known sheep-bells used in the Isle of Man, and Mr. W. B. Gardner tells me of the perhaps stranger fact that sheep-bells are not used in Scotland.

---

[1] See illustration.
[2] Joyce, *op cit*.

SCOTLAND

MY HIGHLAND SHEPHERD FRIENDS

By Kate Henry-Anderson, 1909

On the hills above a small Highland town famous in history, there lived some years ago an old shepherd of the class now so rare. He was eighty-four years of age, hale and strong, with undimmed eyes and unclouded mind. In all his long life he had never been ten miles from the hamlet where he had a cottage, and had spent his days, and many nights, on the moors that stretch beyond Amulree. Here in the long summer light you might find him with his two dogs, one a black-and-tan collie, the other a grey-and-white shaggy West Highland sheep-dog;[1] wise beasts that did all but speak. He carried in his "pouch," or shepherd's bag, an old, much-worn calf-bound Bible, and his peaceful leisure was well filled to him by the rhythm and grandeur of the Psalms of the shepherd king who ages before watched his father's flocks. The old man had been just such another youth, "ruddy, and of a fair countenance"; but glory, its perils and temptations, had passed him by, and left his soul free from the world's dark stain. The secret, sweet influence of the Pleiads, the glory of sun, moon, and stars, the miracles of the seasons, and Nature's magical beauty, had wrought in him a simple faith, a serenity of mind and heart that were plainly written in his clear, gentle eyes and on his noble brow, that heritage of the true Scotch peasant. Fresh-coloured, tall and spare, with the muscular leanness that betokens a hardy outdoor life, only the snow-white hair and the many fine wrinkles of his face showed his great age. He was a mine of tradition and folk-lore, and weather-wise to an extent that set his forecast far above the laird's barometer. I asked him once if he had never wished for a wider life, a sight of the unknown world that lay beyond the hamlet. "Na, na," he answered, with a

[1] A variety of the collie, though called simply sheep-dog.—[*Author's Note.*]

gentle smile. "I hae ma sheep an' the hills, and the Psalms o' David dinna fail me. I want naething better. I'm never weary wi' them to read." More than most men did he realise that the Lord was his Shepherd. His soul has now gone to its eternal home, and an old Highland custom has set on the grey stone above the grave his humble calling after his name, thus: "Alastair Mackenzie, Shepherd, who died at ——, aged 92.

The Lord is my shepherd."

In the beautiful county of Inverness lives a shepherd, who made once, and never again, an effort to live in a great city. He is a sturdy man in the prime of life, brown-bearded and grey-eyed, with the curious gentle look which may be seen in the eyes of wild creatures. We had missed him from the hills, and were told that Donald had gone "awa tae Glesca." But the end of the summer saw his return to the purple moorland.—He said, "I was sae lonely an' ma hairt was sair. I was wae for ma sheep an' the hills. I couldna bear tramp, trampin' thae grey streets, wi ne'er a kent face tae greet me; an' the dirt an' the noise. I was clean demented. So I just cam' awa back, an' I'll no gang ony mair til the toun."

One of the shepherds in Inverness-shire sixty years ago was a bard, and composed and sang his own songs. He would improvise for the delight of the village, and has left a book of Gaelic songs with *sol-fa* tunes. I know his old son Donald, who is eighty years of age. The shepherd of the North used always to wear the "Tam-o'-shanter or bonnet," and a huge plaid curiously folded. In this he slept on the heather very warmly. They often knitted their own hose, and I know one who did so for forty years. Very weather-wise and full of old tales and traditions are these men, and they love the Bible or "the Book."

Another of my shepherd friends is Callum Macpherson. He comes from the far North in Argyllshire, near Loch Shiel. He is of a very old family—"hundreds of years old," he said—a fine man, full of Celtic feeling and imagination, very handsome, with grey eyes and fresh complexion—a true Highlander of my favourite type. I understand the Scot peasantry and the Highland shepherds because I love them, and I find in them the simplicity which is that of "the little child" of the Bible. They have warm hearts of fire and tenderness compounded, and you must be bred among them, meet them, chat with them, and trust them, then the rich treasures of their nature are yours. They are "leal," and never fail you in joy or sorrow. I love to speak of them to those who love to hear.

HARDSHIPS OF SHEPHERD LIFE IN THE HIGHLANDS

By ALEXANDER INNES SHAND, 1905

Not a few of my most enjoyable days have been passed in the company of keepers and simple-minded hill shepherds. They are intelligent, companionable, and instructively conversable when they come to know you well. I have made friends with sundry Highland shepherds, and have a great regard for them, and much sympathy with their hard and solitary lives. . . . Take them all in all they are an honest and self-respecting set of men. Many a weary league from the kirk, their Sunday reading is often the Bible and *Pilgrim's Progress*. The shepherd with his trials and troubles is naturally short in the temper. If he is misanthropic, it is because he so seldom sets eyes on a fellow-creature; but only take him in the right way, and he is the most kindly of hosts and the most friendly of companions. . . .

As a rule the shepherds hasten to be married; but imagine the lot of the celibate, with no company but his collies. His evensong as he goes home in the gloaming is the scream of the eagle or the croak of the raven; and through the nights those dogs of his are baying the moon or answering the challenge of the prowling fox.

Weary and soaked to the skin, he has to do his own cooking, and as he has neither leisure nor energy to shift his clothes, no wonder rheumatism steals upon him early. He knows the lie of the land well ; but many a time when belated in darkness or mists, he has to sleep out in some cleft of the rock, on a couch of damp heather shoots, with a plaid for a coverlet. He is answerable for the sheep, which are periodically mustered and numbered.

Reading the weather like a book, in late autumn he sees the sign of a "breeding storm," and whistling to his dogs he wanders forth to lead back the sheep from the heights to the hollows. . . . The storm bursts and the rain descends in torrents ; all the more reason for the shepherd going forward, for he knows that on the morrow there will have been drownings in the strath, and that eagles and ravens will be battening on the "braxy." He does what may be done before darkness settles down, and then, if it is possible, he would get back to his fireless fireside. But each burn and rill is rising in spate, and the stream from which he fills his water-butt is half-breast high and raging furiously when he gropes his way to the post that marks the ford and the stepping-stones. Within gunshot of supper and the box-bed, he may have to curl up in the moss-flow, with his whimpering dogs, famished and shivering.

Yet this is a trifle to being abroad in the winter blizzards, when the flock may be smothered in the snow-drifts. The bitter wind pierces through the thickest clothing, and he is likely enough to get lost in the blinding snowflakes. A slip on the rocks may sprain an ankle, or, treading carefully as he will, he may fall into a treacherous snow-wreath. Once caught to the armpits, there is slight chance of extraction. All these things considered, it is wonderful that the casualties come so seldom, and that, save in exceptional cases and in the lambing season, so few of the sheep are missing.

These sheep are extraordinarily hardy, and seldom succumb to anything but suffocation. They wear warm under-vests of close wool, with shaggy overcoats as impervious as Irish frieze. They can exist for days on starvation fare, and like the deer have an instinct for scraping among the snow, where they are likely to get at the coarse but nutritious herbage. When the shepherd's strength is taxed to the utmost is in such a storm as is described on Exmoor in *Lorna Doone*, and in the Highlands he has to go farther afield than Jan Rid, to dig into drifts and save the survivors.[1]

The shepherd has other enemies to fight than the snow and rain floods. I do not believe foxes or eagles do much harm to the old sheep, but they are terribly destructive in the lambing season, all the more since the extension of the deer forests. The eagles have been generally strictly preserved, which is gratifying from the picturesque point of view; but if a sheep is crippled or ailing the eagle is always on the look-out, guided by the ravens and hooded crows. For the eagle is the most voracious of gluttons, and the best chance for the shepherd to take his revenge is when he weathers on a bird gorged to the beak with drowned mutton. Then the prince of the air and the mountains may be knocked senseless with the staff. It is not so very easy to circumvent the fleet and wily fox, who does infinitely more harm. He has his lair in the recesses of the half-impregnable cairn, and laughs at the comparative lumbering of the swiftest collies, and is only to be forced from his hold by varmint terriers. Consequently none of his rare visitors is more welcome to the shepherd than the professional fox-hunter with his mixed pack. With the "tail" of his professional dogs come keepers and gillies, each with his own canine attendants, and then dens are stormed and there may be merciless slaughter.

[1] *Pictorial Half-Hours* (1851) has the following :—" In the wild mountain districts of Scotland the shepherd has frequently appalling dangers to undergo from the terrible storms that sometimes desolate those exposed regions. Hogg has given us accounts of several most fearful ones; in one he was himself a participator, and bravely incurred no small hazard in fulfilling his duty. Seventeen of his brother shepherds and an innumerable number of sheep perished on that single occasion."—[*Author's Note.*]

MR. JAMES GARDNER

JAMES GARDNER, SHEPHERD AND FAMOUS COLLIE-DOG TRAINER

An Appreciation by a Grateful Pupil, 1909

Mr. James Gardner belonged to one of the oldest shepherd families in Scotland. He was born in the Upper Ward of Lanarkshire, at a place called Todholes, in the year 1840, and died at North Cobbinshaw, Midlothian, in 1900. He was deeply devoted to his calling, and would have regarded as an insult any suggestion or offer of social advancement which would have entailed his leaving the shepherd world. A true mountaineer, he retained a lofty independence, nor could he tolerate any of the forms of cant and hypocrisy. He was in the best sense a God-fearing man, but made no parade of religion. It was said of him at his death by one who knew him thoroughly: " A truer and more generous man than

James Gardner I have not known; he was quite incapable of a mean action." This very accurately summed up the character of the man. In company he was the very soul of good fellowship, his conversational powers being quite unique. Education and social opportunities would have made him a great orator; the want of these gave him place and power in a humbler and purer world. "Give me," he would say with fine rustic eloquence, "the warmth of my home and the devotion of my dogs, and I will ask none of the vain and fleeting pleasures of the gay world." When he died his dogs mourned his departure with all the pathos of loving and broken-hearted children.

James Gardner married when he was in his eighteenth year, his bride, Mary Black, being in her seventeenth. They had four children, two sons and two daughters. His death was the result of an accident which happened when he was directing the demolition of an old building, and his wife survived him only a year. Much of an abstract nature has been written of the shepherd and his calling, but an experience of Mr. and Mrs. James Gardner's hospitality revealed in a memorable way all the sweetness and charm of life amongst the hills. Their guests were made to feel that warm hearts can convert small cots into great palaces.

JAMES GARDNER—ANOTHER ACCOUNT

By the Rev. Hugh Young, 1910

It was in the spring of 1884 that I first became acquainted with Mr. James Gardner, of North Cobbinshaw, and quickly formed a friendship, close and intimate, which continued unbroken till his death. Daily intercourse with him soon made apparent his fine and sterling qualities as a man, and his unmatched skill and ability as a shepherd. Among my first impressions of him which remained and deepened as the years went on, were his genial nature, his kindly heart, his fidelity to duty, his intense interest in all things human, and his fine sympathy with every living thing in Nature, and, above all, in the flocks of sheep under his charge. He was possessed of a rich fund of folk-lore, as well as a varied, accurate, and extensive knowledge of all animals he came in contact with. To me it was a sincere pleasure to hear him talk by his own fireside of the people of his acquaintance, and with the aid of a bright and vivid imagination describe in glowing language the humorous incidents of his shepherd life. Time and again was this done with unfailing interest, for the fund of stories was inexhaustible, and his powers of description brilliant. I remember well and gratefully his kindly interest in myself when, coming to the district as a clergyman, a stranger, he took me in, not only to his home but to his heart, and not a little encouragement did I receive from him when I observed that he seemed to regard my work of a pastor as being carried on very much on the lines of his own, although admittedly in different spheres, and I had always the feeling that his inward criticism of myself took the same direction ; his ideal of a shepherd and his duties was high, and it was the strenuous effort of his life to attain it. His fidelity to and care over his flock were unapproachable, and his knowledge of sheep and shepherd duties could not be surpassed. Often when discoursing to my people on the Great and Good Shepherd have I had Mr. Gardner in my mind, and have drawn from the admirable, I might say perfect, manner in which he tended his flock my best points and aptest illustrations.

But even still more remarkable than his wonderful ability as a shepherd, was the extraordinary mastery he had over his dogs, as seen in his skill in training and managing them. It was extremely interesting to see him working them on the hill-side, or indeed anywhere. Every turn, every move seemed perfect ; the reasons for this were probably the close intimacy and the friendly understanding that existed between the parties concerned, the master and his servants, and also the fact that he never struck his dogs ; he had always a dread of doing so, no matter how much they were at fault, or how they had disobeyed. This lesson he had learnt, he told me shortly before his death, when he was a young man. He said that he was one day gathering a lot of sheep near the farmhouse of his employer. The dog he was then working had offended him, and in a little anger he threw his crook, not intending to strike, but unfortunately it did. Ever after when he was working at the same place the dog seemed to become paralysed and useless. This was a lesson, he said, he never forgot, and from that day he would never strike a dog. Many are the interesting stories that might be told of Mr. Gardner and his dogs, to show the latter's sagacity, intelligence, and ability in working, as well as their master's command over them. One or two may suffice.

On one occasion when he and I were walking leisurely along the road leading through the farm, Rasp, one of the best dogs he ever had, was working alone around a little hill at a considerable distance. He happened to say in conversation that if he only would say to her, "Ye've left one," she would run away back at once for it. Presently she appeared. He very quietly repeated the words, and away she went to get the supposed left sheep. "Ah," he said, "it will not do to deceive her; I must call her back at once." He did so, and again in an instant she appeared round the brow of the hill, looking up, and apparently wondering what this unprecedented action on the part of her master was about, and little dreaming that it was all done to show off her own splendid qualities and accomplishments. On another occasion, shortly before his death, I was passing the farm; where Mr. Gardner was working with the sheep, in the heart of the "bughts," there was a small group of sheep standing by the barn door, which he wanted quietly brought round to him. He gave Turk, a big, strong, rather impetuous dog, the proper word of command. Turk seemed to understand exactly what was wanted, for in obeying he did little more than paw the ground with his forefeet, and on this slight movement the sheep went quietly trotting round to the very spot where his master was working. I said that was very beautiful. "Yes," he replied, "Turk has done very well just now."

That Mr. Gardner was a sincere and deeply religious man I have not the least doubt. When he could, he attended regularly the services of our church, and a very intelligent and devout worshipper he was. Living far from the church, we had a custom for many a year of having a short service in his own house every Sunday evening. Often, when I went in for the purpose, have I found him reading his Bible, as if to prepare his mind for the sacred duties to follow. His end was sudden, but he died at his post, thinking not so much of himself as of the safety of others, and particularly of the welfare of his flock, ever so dear to his heart, and which he tended with so great a devotion.

SAYINGS ON DOGS, TAKEN FROM THE CONVERSATION OF JAMES GARDNER

In every case a great dog bears a deep resemblance to his master.

I have never known a deceitful man to have a faithful dog.

In training, you have to guide instinct by a superior and kindly wisdom, being always careful not to blunt the genius of the pupil by over-direction.

When a dog bites a man, that man is sorely in need of chastisement.

The higher type of collie is easily spoilt, the lower type can scarcely be rendered useless even by the greatest novice.

The shepherd who creates work to train his dog is unworthy of his calling.

A dog without a "strong eye" has no claim on my patience.

The leading, not the commanding faculty, is the strongest quality of a great dog.

A shepherd's dogs should all be recognised as members of the family. Such an arrangement makes life much fuller and sweeter.

Shepherd and Flock

When my dog wakes from a dream, I know from his look that I have been present in his dream.

Not one of the great dogs of history was ever thrashed into obedience.

No insult would wound me deeper than a look of distrust from one of my dogs.

Prove yourself worth dying for, and your dog, if need be, will cheerfully make sacrifice.

The noblest lessons in truth, sacrifice, and duty I have got from my dogs.

Base-minded men work for money; my dogs work because service is their pleasure.

The moral superiority of my dogs often makes me feel ashamed; they are so grateful for every touch of kindness.

When I am taken away, my dogs will deeply mourn my departure. Why should I not then mourn their death? for, bear in mind, they have done much more for me than I have done for them.

ON SHEPHERDS AND SHEPHERDING IN SKYE

By Alexander Smith, 1865

The pastoral life is more interesting than the agricultural, inasmuch as it deals with a higher order of being; for I suppose — apart from considerations of profit — a couchant ewe, with her young one at her side, or a ram, "with wreathed horns superb," cropping the herbage, is a more pleasing object to the æsthetic sense than a field of mangold-wurzel, flourishing ever so gloriously. The shepherd inhabits a mountain country, lives more completely in the open air, and is acquainted with all the phenomena of storm and calm, the thunder-smoke coiling in the wind, the hawk hanging stationary in the breathless blue. He knows the faces of the hills, recognises the voices of the torrents as if they were children of his own, can unknit their intricate melody as he lies with his dog beside him on the warm slope at noon, separating tone from tone, and giving this to rude crag and that to pebbly bottom. From long intercourse every member of his flock wears to his eye its special individuality. Sheep-farming is a picturesque occupation ; and I think a multitude of sheep descending a hill-side, now outspreading in bleating leisure, now huddling together in the haste of fear — the dogs, urged more by sagacity than by the shepherd's voice, flying along the edges, turning, guiding, changing the shape of the mass—one of the prettiest sights in the world.

AT MR. M'IAN'S FARM.—THE SHEPHERDS AT DINNER

The shepherds, the shepherds' dogs, and the domestic servants, dined in the large kitchen. The kitchen was the most picturesque apartment in the house. There was a huge dresser near the small dusty window ; in a dark corner stood a great cupboard in which crockery was stowed away. The walls and rafters were black with peat-smoke. Dogs were continually sleeping on the floor with their heads resting on their outstretched paws ; and from a frequent start and whine you knew that in dream they were chasing a flock of sheep along the steep hill-side, their masters shouting out orders to them from the valley beneath. The fleeces of sheep which had been found dead on the mountains were nailed on the walls to dry. Braxy hams [1] were suspended from the roof ; strings of fish were hanging above the fireplace. The door was almost continually open, for by the door light mainly entered. Amid a savoury steam of broth and potatos, the shepherds and domestic servants drew in long backless forms to the table, and dined, innocent of knife and fork, the dogs snapping and snarling among their legs ; and when the meal was over the dogs licked the platters.

[1] Cured flesh of sheep which died, but were not butchered. *Braxy* is also a disease among sheep.—(LAUCHLAN MACLEAN WATT.)

Shepherd and Flock

THE PRINCIPAL SHEPHERD

John Kelly was M'Ian's principal shepherd—a swarthy fellow, of Irish descent I fancy, and of infinite wind, endurance, and capacity of drinking whisky. He was a solitary creature, irascible in the extreme ; he crossed and recrossed the farm, I should think, some dozen times every day, and was never seen at church or market without his dog. With his dog only was John Kelly intimate and on perfectly confidential terms. I often wondered what were his thoughts as he wandered through the glens at early morning and saw the fiery mists upstreaming from the shoulders of Blaavin ; or when he sat on a sunny knoll at noon, smoking a black wooden pipe, and watching his dog bringing a flock of sheep down the opposite hill-side. Whatever they were, John kept them strictly to himself.

C. Reid, Wishaw, N.B. *Copyright.*
CALLED TO THE TROUGHS

THE BLACKFACE BREED

By D. MACPHERSON, 1909

In Scotland the sheep are mostly blackfaced, not necessarily black in their wool. Black sheep are to be found in all breeds, but chiefly among the blackface breed. It cannot be said that the sheep we call black are pure black; they generally have a tinge of brown, and in some cases a mixture of grey. Many years ago I wore a suit made of black wool; this needed no dyeing. The practice of the shepherd to carry feeding in his plaid or tartan in the severe weather of spring, is becoming obsolete; they take milk with them to the hills for weakly lambs. The sheep in the fields are at feeding-times (for hand-feeding in boxes) called to the troughs, and the shepherds have their special calls, to which a ready response is made.

DEER EXPELLED BY SHEEP

From The Table Book of William Hone, 1827

A note to a poem, " The last Deer of Beann Doran," by John Hay Allan, relates that in former times the barony of Glen Urcha was celebrated for the number and superior race of its deer. When the chieftains relinquished their ancient character and their ancient sports, and sheep were introduced into the country, the want of protection and the antipathy of the deer to the intruding animals gradually expelled the deer from the face of the country, and obliged them to retire to the most remote recesses of the mountains. Contracted in their haunts from " corrie " to " corrie," the deer of Glen Urcha at length wholly confined themselves to Beann Doran, a mountain near the solitary wilds of Glen Lyon, and the vast and desolate mosses which stretch from the Black Mount to Loch Rannoch. In this retreat they continued for several years. Their dwelling was in a lonely corrie at the back of the hill, and they were never seen in the surrounding country except in the deepest severity of winter, when, forced by hunger and the snow, a straggler ventured down into the straths. But the hostility which had banished them from their ancient range did not respect their last retreat. The sheep continually encroached upon their bounds and contracted their resources of subsistence. Deprived of the protection of the laird, those which ventured from their haunt were cut off without mercy or fair chase; while want of range and the inroads of poachers continually diminished their numbers, till at length the race became extinct. . . .

The same cause which has extirpated the deer from Glen Urcha has similarly acted in most parts of the Highlands. Wherever the sheep appear their numbers begin to decrease, and at length they become totally extinct. The reason of this apparently singular consequence is the closeness with which the sheep feed, and which, where they abound, so consumes the pasturage as not to leave sufficient for the deer; still more is it owing to the unconquerable antipathy which these animals have for the former. This dislike is so great that they cannot endure the smell of their wool, and never mix with them in the most remote situations, or where there is the most ample pasturage for both. They have no abhorrence of this kind for cattle, but where large herds of these are kept will feed and lie among the stirks and steers with the greatest familiarity.

SCOTTISH SHEPHERDS OF THE SIXTEENTH CENTURY

By JAMES TAYLOR, D.D., 1859

A graphic and very interesting description of the manners and occupations of Scottish pastoral life at this period is given by the author of the *Complaint of Scotland*.[1] The shepherds are represented as wearing hoods which covered their heads and shoulders and conveniently admitted the additional envelope of the plaid. They amuse themselves with the buckhorn and corn-pipe, while their flocks graze along the "banks and braes" and dry hills. About breakfast-time they are joined by their wives and daughters, who bring their food and prepare them a seat, spreading the soft yellow moss of a lea ridge with rushes, sedges, and meadow-wort, or queen of the meadow. The food of which they partake consists principally of

---
[1] Attributed to "Sir" James Inglis; rare. Published 1548.—[*Author's Note.*]

various preparations of milk, all of them still well known in Scotland. "They make good cheer of every sort of milk, baith of cow milk and ewe milk, sweet milk and sour milk, curds and whey, sour kitts,[1] fresh butter and salt butter, reyme,[2] flotwhey,[3] green cheese and kirn milk.[4] They had na bread but rye cakes and fustean[5] scones, made of flour." Every shepherd is represented as carrying a spoon in the "lug (ear) of his bonnet"—an extremely characteristic circumstance, for even down to a very recent period not only shepherds, but reapers and peat-diggers, frequently provided themselves with spoons, which they carried about with them in the manner described. On the conclusion of their simple meal the shepherds amused each other by relating in turn tales or stories in prose and verse, and their wives "sang sweet melodious sangs." The entertainment at length terminated in a general dance to the music of eight different kinds of instruments, after which the shepherds collected their flocks and drove them tumultuously to the folds. This simple representation is accurately copied from nature, and Dr. Leyden says the original might still be seen in his day in some of the wild pastoral districts of Scotland. As the flocks of sheep, after grazing some hours, are always disposed to rest in the sunny days of summer, basking themselves on some dry acclivity, a concourse of shepherds for a social meal enlivened with songs and stories, and occasionally diversified by a dance, was by no means an uncommon incident.[6]

[1] Clouted cream. Kit is a small kind of wooden vessel, hooped and staved.
[2] Cream.
[3] A common dish in the pastoral districts of Scotland, formed by boiling the whey, after it is expressed from the cheese curds, with a little meal and milk, when a species of very soft curd floats at the top.
[4] Churn milk, fustean.
[5] Fustean signifies soft, elastic. Hence "fustean scones" are cakes leavened or puffed up. Scones are cakes made of wheat, rye, or barley meal. This term is never applied to bread made of oats.
[6] Leyden's Preliminary Dissertation to the *Complaint of Scotland*, p. 128.

THE MILKING OF EWES

By The Author

The mention of ewe milk in the preceding article is of interest. Mr. H. M. Doughty, in *Chronicles of Theberton* (1910),[1] writes that "Ewes were milked for the dairy, as is still done in Holland, and produced more for the grass they consumed than cows."

> Five ewes to a cow, make a proof by a score,
> Shall double thy dairy, or trust me no more.
> Thomas Tusser,
> *500 Points of Good Husbandrie* (1560).

William Bingley, writing in 1816, says: "From the milk of the Cheviot sheep great quantities of cheese are made, which is sold at a low price. This, when three or four years old, becomes very pungent, and is in considerable esteem for the table." And *The Antiquary's Portfolio* (1825) has: "Formerly in many parts of England cheese was made from the milk of the ewe, and the ewes, to the injury of the lambs, were milked regularly, as described in *The Odyssey*:

> "He next betakes him to his evening cares,
> And sitting down to milk his ewes prepares."

With regard to Ireland, it is a significant fact that the ancient pagan Irish name for St. Bridget's Feast on February 1st, which was anciently reckoned as the first day of spring, was Ewe-milk or the Ewe-milking. In Scotland during the season that the ewes were milked the bught door was always carefully shut at even, to keep out the witches and fairies, who would otherwise dance in it all night.[2]

In Jane Elliot's famous and touching song founded on *The Lament for Flodden*, there are several verses which are to the point here:

---

[1] Messrs. Macmillan & Co.
[2] James Hogg, *Mountain Bard*.

Shepherds of Britain

I've heard them lilting
At the ewe-milking,
Lasses a' lilting
  Before dawn of day;
But now they are moaning
On ilka green loaning ;
The Flowers of the Forest
Are a' wede away.

At bughts in the morning
Nae blithe lads are scorning ;
Lasses are lonely
  And dowie and wae ;
Nae daffing, nae gabbing,
But sighing and sabbing ;
Ilk ane lifts her leglin
And hies her away.

In har'st at the shearing
Nae youths now are jeering ;
Bandsters are runkled,
  And lyart or grey ;
At fair or at preaching,
Nae wooing, nae fleeching :
The Flowers of the Forest
Are a' wede away.

. . . . .

We'll hear nae mair lilting
At the ewe-milking,
Women and bairns are
  Heartless and wae ;
Sighing and moaning
On ilka green loaning,
The Flowers of the Forest
Are a' wede away.[1]

---

[1] *The Illustrated Book of Scottish Song* gives the following note :—" The Flowers of the Forest are the young men of the district of Selkirkshire and Peeblesshire, anciently known as 'the Forest.' The song is founded by the author on an older composition deploring the loss of the Scotch at Flodden Field, of which all has been lost except two or three lines." Jane Elliot of Minto was sister of Sir Gilbert Elliot, the author of the beautiful pastoral song, " My sheep I neglected, I lost my sheep hook."—[*Author's Note.*]

SHETLAND AND ORKNEY ISLES

THE WILD SHEEP OF SHETLAND

By Dr. S. Hibbert, 1822

The tenants of the *Scatholds*[1] were the wild sheep of the country, celebrated for their small size, and known by naturalists under the name of the *Ovis cauda brevi*, that at the present day range among the mountains of modern Scandinavia and Russia. In very few places are the Shetland sheep mixed with a Northumberland breed. . . . In summer they collect from the pastures that kind of food which the natives still designate by the ancient Scandinavian term of *lubba*.[2] The sea also affords provision, which, in the severer months of the year, prompts them, upon the ebbing of the water, to flee to the shore, where they remain feeding on marine plants until the flow of the tide; they then return to the hills. . . . The sheep are allowed to run wild among the hills, herding and housing being almost unknown in Shetland. There is an old law, that was probably introduced by the Scotch settlers, ordering that every scathold have a sufficient herd, and that builing,[3] punding,[4] and herding be used in a lawful

---

[1] "The whole interior of the country consists of the unimproved and generally undivided common or scathold. These large tracts of country are made up of peat-moss, lochs, or bare ground, which has been robbed of soil by 'scalping.' They are clothed, where vegetation can exist, with heaths and coarse grasses, interspersed with *carices* and rushes. In the damper and poorer spots moss is the only plant to be found."—Cowie, 1871.—[*Author's Note.*]

[2] Coarse grass of any kind (*Juncus squarrosus*); Danish, *lubben*; Icelandic, *lubbe* = coarse grass.—Edmondston's *Glossary of Shetland and Orkney Words*.—[*Author's Note.*]

[3] To put in a quiet place for the night. Icelandic, *bola*, to lie down.—Edmondston's *Glossary*.—[*Author's Note.*]

[4] Though Edmondston in his *Shetland Glossary* gives *pund* (Anglo-Saxon, *pynd-an*, to shut up, to enclose) he does not mention punding, but the latter word is in common use in Shetland. Hibbert seems to use it in the sense of putting the sheep into punds or crues before marking or rueing them; but the general use of the word is applied to stray sheep, which are put into punds or enclosures or even office-houses, and are kept there until relieved by their owners, who may have to pay fines to the persons who punded them.—Rev. T. Mathewson.

In England we formerly used the word pinfold in the same way. A pinfold or enclosure for cattle or strayed animals. And the obsolete form "pindar" was from the same origin (1523, Fitzherb. *Husb.* § 148). "Then cometh the pynder and taketh hym and putteth hym in the pynfolde." The pinder or pounder, *i.e.* impounder, was an officer of the manor, whose duty it was to impound stray beasts (1632, Title, *The Pinder of Wakefield: being the merry history of George A. Greene, the lusty Pinder of the North*).—N.E.D.—[*Author's Note.*]

way, before or a little after sunsetting, but the regulation has not, for a long time, been enforced. On the contrary, the sheep are almost to be regarded as in a state of nature, since they range at large over the scatholds during the whole of the year.

Whenever it is requisite to catch any sheep, they are hunted down with dogs trained for the purpose, which Wallace, the historian of Orkney, describes as a sport "strange and delectable." The dog bounds after his prey, the flock are immediately alarmed, the poor animal is chased from hill to hill until he falls into the power of his pursuer. . . . Disdaining the use of shears, the wool is torn from the struggling animal's back in the most brutal manner, and if the fleece has not begun to loosen naturally, the operation is attended with most excruciating pain.

SHETLAND SHEEP—ANOTHER ACCOUNT
By Robert Cowie, 1871

The ancient Norse inhabitants of these islands, at first pirates, in course of time became shepherds, but they never appeared to excel in agriculture. In Shetland the inland landscapes are comparatively tame and monotonous, the rock scenery is always interesting, and often truly grand and magnificent. Look to the land, and you behold hills beyond hills, the brown sward of each dappled over with the small species of sheep peculiar to the country, or the better-known Shetland pony, browsing on the coarse grass or the more tender shoots of the young heather. The sheep find both food and shelter for themselves as best they can all the year round. They are most valuable, not only as food and articles of sale, but also as affording the women raw material, out of which to produce their far-famed hosiery. . . . The native Shetland sheep is of the same species as those which run wild in Northern Russia and Scandinavia. Besides the short tail, they are characterised by small size, fine wool, and short horns. They run wild in the scatholds, are never housed,

Shepherd and Flock

herded, or fed by hand, and those of different owners are distinguished by characteristic slits or holes in the ears. When an individual sheep in the flock is wanted by its owner, it is hunted down by a dog. In colour these sheep are white, black, spotted black and white, grey, or of a peculiar brownish shade, termed by the natives *muirid*.[1] . . . Birds of prey often prove destructive to the young lambs. Their most formidable enemies are ravens and hooded crows. Even the black-backed gull sometimes attacks them. But the most blood-thirsty of all are eagles, the extent of whose ravages is only limited by their numbers, which fortunately are small.

In the winter, particularly when grass is more than usually scanty, both sheep and ponies frequently feed on seaweed ; they may be observed with wonderful sagacity approaching the most accessible shore when the tide begins to fall, and leaving it as high water sets in.

Before the fishing became an object of so much regard, far greater attention was paid to the raising of sheep than has ever been since. This is shown by the large proportion of the old county acts which are devoted to the regulation of the pastures.

SHEPHERDING IN SHETLAND AND ORKNEY

By the Rev. T. MATHEWSON, 1909

During the last century these islands have become much more Scotch in their manners and customs. Shepherds from the south have come to settle, and consequently their modes are of a southern rather than a northern manner. Among the crofters there is no such thing as shepherds and shepherding in the usual meaning of the term ; the sheep roam at large, and are only looked after in the lambing season, when they are being "rued,"[2] or driven off to be sold.[3] When I was a boy the crofters

---
[1] *Muirid* or *murrit*—moor-coloured. We note that Cowie talks of the brown sward of the hills.—[*Author's Note.*]
[2] Or plucked.
[3] Though the shepherding is somewhat casual, the shepherd loves his flock. Mr. S. R. Tatham writes that "when fishing in Orkney he has the services of an old

generally "builed" the sheep some distance from their houses, so that they might not feed upon the growing corn, but of late years the crofts have got better dykes and railings, and that is no longer necessary. The shepherds have been mostly from Scotland, and native ones have been few. They follow the sheep; indeed the regular expression is to ca' (drive) the sheep. The collie is the dog generally used. By some instinctive power the sheep seem to know when it is ebb tide, and rush from the hills, or from wherever they are, and take their fill of the seaweed. When we were children we were told by our elders that there was a worm in each of the sheep's forefeet, and that this worm warned them when it was an ebb. I remember when a boy cutting out this worm after the foot was boiled, but I have been told that others cut them out before cooking them.

In Dr. Hibbert's day (early last century) "the skins of the Shetland sheep were in requisition for the purpose of affording the fishermen a sort of surtout that covers his common dress." In my younger days the skin was taken off whole from the animal, and used as a bag for holding meal. It was called a buggie, and the act of taking off the skin in this fashion was called buggie-flaying. James Macdonald in his *Place Names in Strathbogie* says that "the root of the name of the river Bogie which runs close by my rectory is 'bolg,' a sack or bag, generally a leathern bag corresponding to what in Shetland is called a 'buggie,' that is a bag made from a sheep's skin removed from the animal, from the neck downwards, so that the skin is left almost entire." Sieves and weights for sifting and holding corn were made from sheepskins stretched on hoops, the holes of the former being bored by a red-hot wire.

<small>shepherd to row the boat, and that he usually disappears for an hour in the middle of the day to visit his flock." On one occasion he asked for a "day off," as he had to drive the sheep to market, and on the following morning, as he seemed to be depressed, Mr. Tatham asked him if he did not get fond of his sheep and feel parting with them on such occasions. His answer was, "Surely, surely."—[*Author's Note.*]</small>

BREEDS OF SHEEP IN SHETLAND AND ORKNEY

By JAMES JOHNSTON, 1909.

In 1908 the number of sheep in Orkney was 27,504; in Shetland, 133,955.

Most of the sheep are clipped in the ordinary way in these islands, but some of the native sheep are "rued" in Shetland. This is done in July, when the wool is coming off. In Orkney there are Cheviots, blackface, Leicesters, and a few native sheep in North Ronaldshay, which pasture in common along the seashore and eat the seaweed. In Shetland there are native sheep, blackface, Cheviots, Leicesters, and half-breeds. More than half are natives. The wool of the latter is fine, and sells for 1s. 10d. to 2s. 6d. a lb., and is used for making the famous shawls, etc. The blackface are next in number.

RARER PHASES OF SHEEP CULTURE AND CHARACTER

RARER PHASES OF SHEEP CULTURE AND CHARACTER

SHEEP LED BY THE SHEPHERD

By The Author

This is by no means such a rare occurrence in England as some think. Many of the Sussex shepherds tell me that they lead their sheep, especially on the Downs. But much depends on whether the dog is trained to the method or not.

William Ellis, in his *Shepherd's Sure Guide* (1749), quotes "a very ancient writer," whose account of shepherding in France bears on our subject and seems quaint enough to be worth recording. "'The shepherd,' says he, 'shall order and govern his flock with great gentleness, as is most requisite for all herds; who must rather be and show themselves leaders and guides of their beasts than lords. Guiding them to the field, he must always go before them to hinder and keep them back from running into fields where they might feed upon evil and hurtful grass; and especially such grounds as wherein water useth to stand. . . . He shall,' says he, 'rather keep a white dog than one of any other colour to follow his sheep; and he himself must also be apparel'd in white; because that the sheep are naturally so inclined to fear, as that and if they see but a beast of any other colour, they doubt presently, that it is the wolf which cometh to devour them. This dog must have a collar of iron about his neck, beset with good sharp points of nails, to the end that he may the more cheerfully fight with the wolf. And if it happen that his sheep be scattered, to call them in and bring them together again, whether it be for keeping them out of harm, or to cause them to know his call; he must whoop and whistle after them, threatening them with his sheep-crook or else setting the dog after them. He must sometimes make them merry, cheering them with songs, or else by his whistle and pipe, for the sheep at the hearing thereof will feed the more hungerly, they will not straggle so far abroad, but they will love him the better.' Hence it is, I suppose, that some shepherds divert themselves with dancing and playing the tabor and pipe, as some do at this time in England, as they did in France when this author wrote."

Michael Drayton (1563-1631) represents the shepherd Melanthus leading his sheep and playing to them:

> When th' evening doth approach I to my bagpipe take,
> And to my grazing flocks such music then I make
> That they forbear to feed : then me a king you see,
> I playing go before, my subjects follow me.

SHEPHERDESSES OF THE SEVENTEENTH CENTURY

By The Author

As time goes on and shepherds become more scarce, shall we not once again have recourse to shepherdesses? Surely among the gentler sex there may always be found suitable guardians of our flocks.

Dorothy Osborne, in one of her delightful letters to Sir William Temple,[1] describes her doings at Chicksands in Bedfordshire, and tells us of real live shepherdesses in 1653. "The heat of the day is spent in reading or working, and at about six or seven o'clock I walk out into a common that lies hard by the house, where a great many young wenches keep sheep and cows, and sit in the shade singing of ballads. I go to them and compare their voices and beauties to some ancient shepherdesses that I have read of, and find a vast difference; but trust me, I think

[1] *Letters from Dorothy Osborne to Sir William Temple*, by Judge Parry.—[*Inserted by permission.*]

these are as innocent as those could be. I talk to them and find they want nothing to make them the happiest people in the world but the knowledge that they are so."

Where was the knitting of earlier days? Sir Philip Sidney in his *Arcadia* did not allow his shepherdesses to be idle while watching their flocks; thus:

> Where sat the young shepherdess knitting, whose voice
> Comforted her hands to work, and her hands kept
> Time to her voice's music.

And a few years later William Brown writes:

> Yonder a shepherdess knits by the springs,
> Her hands still keeping time to what she sings.

AN OLD-TIME LINCOLNSHIRE DROVER

From the *Evening News*, 1908

"Old Bill Thacker" of Gedney, Lincolnshire, is now nearing ninety years of age, and is probably the only drover left of his class in the county. Scores and scores of times, sixty and seventy years ago, Thacker has passed through the eastern counties with large droves of cattle, on his way to Norwich, London, and other markets. The journey to London occupied four to five days, and the drover, or "topsman," as he was known, having got rid of his stock, would walk back into Lincolnshire and pick up another drove, and return to London, Norwich, or some other centre again. A thirty, forty, or even fifty miles' walk was nothing.

It was no uncommon thing for a drover to have the care of five thousand to six thousand sheep, or three hundred or four hundred head of cattle, for ten or twelve days. The great objection of the drovers in those days was not to the roads—difficult to negotiate as they were—but to the old toll-bars, which not only meant the payment of large sums as toll, but also caused irritating delays, the toll-collectors being described as not of the most amiable type.

THE LITTLE NORTHAMPTONSHIRE DROVER

From *The Gentleman's Magazine*, 1797

Having rambled to the junction of the two roads upon Chalk Hill on the sultry morning of July 24, 1797, I rested until a boy, trudging and singing at a great rate, came up to me. "Come along the old road, sir," said he; "it is a mortal sight nearer, and I suppose you are thinking which to take." I found my companion a most famous little chatterer, not much above three feet high, and fifteen years of age. He told me he had been to Smithfield with some sheep; that he went every week, and had thirty miles to walk before night. His frock (smock) was compactly bound up and tied across his shoulders. The straps of his shoes formed a studied cross below the buckles, which he took care to tell me had cost him ninepence in London the Saturday before. Turnpike tickets were stuck in his hatband, noticing the number of sheep he had paid for; and the lash of his whip was twisted round the handle, which he converted into a walking-stick.

I soon found, though so small a being, he was a character of no little consequence upon the road, and he told me any returning chaise or tax-cart would give him a lift for nothing. He was familiar with every one we passed. He wanted no hints to make him loquacious, and thus his busy mind unfolded itself: "Now, sir, do you know, I have a very good master; and he promises if I behave well to make a man of me. When I went to live with him I was a poor, ragged, half-starved parish boy, without father or mother, or never had any as I know of. I have now two better coats than this" (which, by the by, was all one complete shred of darn and patch-work), "and I have a spick-and-span new hat I never had on but Whit-Sunday last, and I am to learn too" (proudly stretching himself and brushing up his eyebrows), "my master says, to write; but he has told me so such a mortal while I fear he will forget it." I asked him if he could read. "Aye, in the Testament. I have almost finished the Gospel according to St. John; and I can repeat the Lord's Prayer and Belief too"—the latter of which he ran over as quick as possible, and asked me if he had missed a word. . . . The *naïve* simplicity with which he delivered himself made him rise rapidly in my good opinion, and as we paced on, he repaid every nod he received with manifold interest. . . . On parting, as I was turning a corner which took me out of sight, he shrilled out, "God bless you!"

THE INFLUENCE OF ENVIRONMENT

By The Author

John Dyer in his poem *The Fleece*, written in 1761, in reference to the Leicester breed of sheep, states that they thrive best on a hilly pasture, consisting uniformly of rich "saponaceous" loam, or marl mixed with clay (not on gravelly soils). To this he adds that the marl, being too cold for the sheep to sleep on in winter (its effect being to cause the sheep to waste), it is necessary for the shepherd

> To sink a trench, and on the hedge-long bank
> Sow frequent sand, with lime, and dark manure,
> Which to the liquid element will yield
> A porous way, a passage to the foe.

The old pastures are the best, as new herbage causes coughing; pasture with a southern aspect, not too shady, must be selected, as bleak weather detracts from the quality of the wool. Sheep can bear no extremes, whether of climate or even of "salubrious food," which

> As sure destroys as famine or the wolf.

The Welsh shepherd of to-day will tell you that his sheep feeding on slopes facing south or west have wool of a much finer texture than those feeding on slopes facing north or east. Arthur Young, in his *Tour in Ireland* (1777), writes of the fine turf producing fine wool, but the quantity did not equal the quality. " To Kildare, crossing the Curragh, so famous for its turf. It is a sheep-walk of above 4000 English acres, forming a more beautiful lawn than the hand of art ever made. Nothing can exceed the extreme

By Habberton Lulham.
ON THE SOUTH DOWNS ABOVE FULKING

"In large pastures shepherds should take care to drive their flocks to the north side, so that they may feed opposite to the south."—PLINY.

softness of the turf, which is of a verdure that charms the eye, and is highly set off by the gentle inequality of surface. The soil is a fine dry loam, on a stony bottom; it is fed by many large flocks, turned out by the occupiers of the adjacent farms, who alone have the right, and pay great rents on that account. It is the only considerable common in the kingdom. The sheep yield very little wool, not more than 3 lb. per fleece, but of a very fine quality."

THE "FLEECY RACHAEL" WEEPING FOR HER CHILDREN

By Alexander Smith, 1865

The most affecting incident of shepherd life is the weaning of the lambs—affecting, because it reveals passions in the fleecy flocks, the manifestation of which we are accustomed to consider ornamental in ourselves. From all the hills, men and dogs drive the flocks down into a fold, or "fank," as it is called here, consisting of several chambers or compartments. Into these compartments the sheep are huddled, and then the separation takes place. The ewes are returned to the mountains, the lambs are driven away to some spot where the pasture is rich and where they are watched day and night. Midnight comes with dews and stars; the lambs are peacefully couched. Suddenly they are restless, ill at ease, goaded by some sore, unknown want, and seem disposed to scatter wildly in every direction; but the shepherds are wary, the dogs swift and sure, and after a little while the perturbation is allayed and they are quiet again. Walk up now to the "fank." The full moon is riding between the hills, filling the glens with lustres and floating mysterious glooms. Listen! You hear it on every side of you, till it dies away in the silence of distance—the fleecy Rachael weeping for her children! The turf walls of the "fank" are in shadow, but something seems to be moving there. As you approach, it disappears with a quick, short bleat and a hurry of tiny hoofs. Wonderful mystery of instinct! Affection, all the more pathetic that it is so wrapt in darkness, hardly knowing its own meaning. For nights and nights the creatures will be found haunting about those turfen walls, seeking the young that have been taken away.

MUTUAL RECOGNITION BY SHEEP AFTER SHEARING

By The Author

In Gilbert White's *Natural History of Selborne* (1789) we read : "After ewes and lambs are shorn, there is great confusion and bleating, neither the dams nor the young being able to distinguish one another as before. This embarrassment seems not so much to arise from the loss of the fleece, which may occasion an alteration in their appearance, as from the defect of that *notus odor*, discriminating each individual personally, which also is confounded by the strong scent of the pitch and tar wherewith they are newly marked ; for the brute creation recognise each other more from the smell than the sight, and in matters of identity and diversity the appeal is much more to their noses than to their eyes. After sheep have been washed there is the same confusion, for the reason given above." In the lambing season the shepherds take advantage of the knowledge that ewes recognise their lambs by scent. William Howitt, writing in 1835, says: "The shepherd has sometimes to find foster-mothers for poor orphans, which is done by clothing them in the skins of the dead lambs of those ewes to which they are consigned." This subtlety is still resorted to by shepherds.

ON PASTURE POISONOUS TO SHEEP

By The Author

Mr. H. Rider Haggard tells us how "Leiston Abbey Farm (in Suffolk) used to be farmed by old monks. Mr. Geaton had laid down twenty-two acres of pasture, but there was some poison on the farm which made it impossible for him to keep sheep. It was a matter of tradition that sheep had always died there, and this was his own experience. In the previous season they had perished even on mustard, to the number sometimes of three a

Rarer Phases of Sheep Culture

night. Once or twice in the course of my travels I have come across farms which were said to be poisonous to sheep; but whether this is so, or they are but temporarily infected with some germ or parasite, is more than I can determine."[1] The original abbey, a mile distant, and built about the year 1182, was found to be "unwholesome and was abandoned; so it would appear that the neighbourhood was not even then blest with health."

Shepherd Stacey, of Lavant, near Chichester, says some pastures are poisonous to sheep at certain times of the year. An insect called flounders infests the root of a plant, and if the sheep eat the infected root they get fat and die. If the sheep eat a certain herb they die from liver rot, or liver-fluke, as it is called.

John Dyer, in *The Fleece* (1761), alludes to "pennygrass and shearwort's poisonous leaf"; and in various counties in England the country-folk call pennywort "sheep-killing." In the Isle of Man *ouw* or marsh-pennywort is well known to be poisonous to sheep. The *Atropa Belladonna* is another of the enemies of the sheep, and has been stated to produce spasms known to shepherds as the leaping ill.

In Bateman's *Great Landowners* (1878) is the following: "Among Lord Dusany's Irish property is one field of a few acres which is remarkable for its fatal effects on all live stock. Horses, if grazed on it, lose their hoofs; stock fed on the hay, if hay be made from it, lose their hoofs, and if the diet be continued they die; if corn or potatos be grown on it, the human animal who eats them loses his nails."

Arthur Young, in his *Tour in Ireland* (1776-1779), records some poisonous pasture in County Sligo, near "Ballasadore." "The mountains nearest to the sea are chiefly stocked with sheep, and further in, with young cattle near the bog. Upon a part of these mountains, of three miles in extent, whatever sheep feed are immediately killed by the staggers, and horses are similarly affected. There is a good deal of limestone in the

[1] *Rural England* (1902). Messrs. Longmans, Green & Co.—[*By permission.*]

neighbourhood, and the land is dry, and [in appearance and in fact] good; it fattens bullocks. Its effect on sheep is attributed to the mines of lead, of which mineral this part of the country is supposed to be full. When they are first affected, if brought down to a salt marsh, they recover immediately."

DEPENDANCE OF SHEEP ON THE WEATHER FOR FOOD AND DRINK

By Richard Jefferies, 1880

Once now and then in the cycle of years there comes a summer which to the hills is almost like a fever to the blood, wasting and drying up with its heat the green things upon which animal life depends, so that drought and famine go hand in hand. The days go by and grow to weeks, the weeks lengthen to months, and still no rain. . . . The shepherds say the mists carry away the rain; certainly it does not come.

Under the beautiful sky and the glorious sun there rises up a pitiful cry the livelong day; it is the quavering bleat of the sheep as their strength slowly ebbs out of them for the lack of food. Green crops and roots fail, the aftermath in the meadows beneath will not grow, week after week "keep" becomes scarcer and more expensive, and there is, in fact, a famine. Of all animals a starved sheep is the most wretched to contemplate, not only because of the angularity of outline, and the cavernous depressions where fat and flesh should be, but because the associations of many generations have given the sheep a peculiar claim upon humanity. They hang entirely on human help. They watch for the shepherd as though he were their father—and when he comes he can do no good, so that there is no more painful spectacle than a fold during a drought upon the hills.

Once upon a time, passing on foot for a distance of some twenty-five miles across these hills and grassy uplands, I could not help comparing the scene to what

Rarer Phases of Sheep Culture

travellers tell us of desert lands and foreign famines. The whole of that long summer's day, as I hastened southwards, eager for the beach and the scent of the sea,

From a painting by R. Westall, R.A. *Engraved by R. M. Meadows, 1813.*
A RAINSTORM
"Happy Britain, that experiences drought so seldom!"

I passed flocks of dying sheep—in the hollows by the way their skeletons were here and there to be seen, the gaunt ribs protruding upwards in the horrible manner that the ribs of dead creatures do. Crowds of flies buzzed in the air. Upon the hurdles perched the crow, bold with over-feasting, and hardly turning to look at me, waiting there till the next lamb should fall and the spirit of the beast go downward. Happy England, that experiences these things so seldom, and even then so locally that barely one in ten hears of or sees them.[1]

THE SNAIL-EATER

By H. L. F. Guermonprez, 1910

It is a popular idea that the excellent flavour of our Southdown mutton is in great measure due to the diet of snails which is said to be indulged in by the melodious bellwether-led flocks browsing on that wonderful stretch of breezy pasture, the South Downs. I am afraid that, like most popular ideas, this has very little foundation in absolute fact, for though snails are undoubtedly at times very abundant on the herbage of our Downs, most of the same species of snails are equally, if not more, plentiful in similar situations in many other parts of Britain. There is only one Downland snail that is at all peculiar to our district; this is the "Carthusian snail" (*Helix carthusiana*), which is practically confined to the South Downland from the Sussex border of Hampshire, through Sussex, and on to Dover in Kent. Some have been found in the Isle of Wight, probably strays, and there is a colony in Norfolk and Suffolk. But this snail is by no means abundant, even on our Downs, and to attribute the excellence of the mutton to it would be a great stretch of imagination. The five other species which may be considered peculiarly "sheep snails" are the "zoned" or "banded snail" (*Helix virgata*), "heath" (*H. ericetorum*), "wrinkled" (*H. caperata*), "Kentish" (*H. cantiana*), and the "acute" (*H. acuta*). All of these are to be found in many

---

[1] In *Chronicles of Theberton*, by H. M. Doughty (1910, Messrs. Macmillan & Co.), we read : " Before turnips and mangolds were known in England, farmers cut down boughs for sheep and cattle." The lack of trees on the Downs is thus an additional evil in time of drought. The author adds : " In winter, when hay ran short, times were hard for the beasts."

" If snow do continue, sheep hardly that fare
Crave mistle and ivy for them to spare."—[*Author's Note.*]

parts of Britain, and in wet weather, more especially when they climb the leaves and culms,[1] they are no doubt consumed by the grass-cropping multitude; but whether this is done by predilection or of necessity, I cannot say. It would certainly appear almost an impossibility for the sheep to avoid eating some, for at certain seasons their number is prodigious, nearly every blade of grass bearing its contingent. In parts of Cornwall the species *Helix acuta* is very abundant, and here the shepherds say that the sheep seek for the snail,[2] browsing on those parts very often at the edge of the cliff, where this snail is more numerous. This may, however, be just as well accounted for by the grass here being longer, as being less accessible, or otherwise more desirable, though it is certainly true that the number of snails there is greater. But if the sheep's instinct works by analogy, their feelings should be an avoidance of a snail diet, for it is by eating a marsh snail, the "dwarf" *Limnaea* (*Limnaea truncatula*), that the liver-fluke trematode (*Distoma hepatica*), a parasitic worm, enters into the organisation of the sheep, and sets up the dreaded "rot." ... Of course, sheep may be good enough naturalists to distinguish between the species, but this *Limnaea* and *Helix acuta* are really, broadly speaking, much alike, both being high-spired. While the dangerous species occurs in damp low ground, or even on the seashore, at the foot of cliffs, the safe one is a dry, high-ground dweller, so the instinct of feeding high and on the edge may be an avoidance one, the poor sheep only falling a victim to the scourge when thoughtless man drives the flock into low, damp meadows, and thus forces them to consume their destroyer. It is stated that over three million animals perished in this country during the winter of 1879-1880 from this complaint. The most common species of "sheep snails" on our Downs is *Helix virgata*, the "zoned" snail, and this is at times so plentiful that one can well understand the popular idea that it has "rained

---

[1] Stems of grasses.
[2] Cp. "Sheep in Cornwall," *supra*, p. 29; and also A. Beckett, *The Spirit of the Downs*, p. 13.

snails." The "zoned" snail is about half an inch in diameter, usually whitish-brown, with a black band following the spiral periphery. The "heath" is rather larger and with many bands; the "wrinkled," smaller and more rugged; the "Kentish," uniform white cream colour, as is also the "Carthusian," which is somewhat smaller; the "acute," very distinct in its cigar, high-spired shape.

THE BONE-EATER

By The Author

Shepherd Smith[1] of Chichester, whose home was at Washington, a hamlet close to the Sussex Downs, tells me of a Southdown sheep that was called the Bone-Eater. When he took his flock to graze at Kingley Vale, this sheep used to wander to a part where stoats and weasels made havoc among the rabbits, and the ground was strewed with bones. She was always to be found here feeding on the bones, on which she throve splendidly. Smith had a thorough-bred collie (bred at Goodwood) which was each day sent to fetch the Bone-Eater and bring her back to the flock. There is a tradition that in the year 895 some Danish "kings," killed in a battle fought near Chichester, were buried in Kingley Vale, a possible reason why this beautiful, though at times gloomy vale might seem especially to suit the fancy of the Bone-Eater.

A Cumberland shepherd says that sheep sometimes kill rabbits and eat them. An informant, writing from Appleby in 1909, describes how he saw three sheep round a young rabbit, which they were tossing over and over with their horns and stamping around in sheep fashion. The poor little thing was squeaking, but too frightened to attempt to escape. He drove the sheep off; and the rabbit, unhurt, was soon able to scamper away to motherland.

[1] For portrait of Shepherd Smith, see p. 250.

The Blind Sheep

An anonymous writer in *Sunday* (1907) tells the following story:—"A flock of sheep, which had just been bought, were being driven to their new home, when the shepherd noticed that one of them was always falling behind the rest and standing still. Every time that it did so, it gave a peculiar plaintive bleat. To his surprise, another of the sheep ran back to walk beside it, until they came up to the rest of the flock. When safely enclosed in their park the shepherd examined the loiterer, and found the poor thing to be quite blind; so that was why its companion came to its assistance when it was bewildered from not seeing where to go." I can record a somewhat similar exhibition of sympathy. On August 17, 1909, a number of lambs were passing my house at 5 A.M. Even the early morning was exceptionally hot, and there was a sad chorus of bleats, an especially piteous one coming from a lamb in the rear. Another lamb was seen to leave its more sturdy companions and run, as it seemed, to give the weary traveller a caress and possibly a word of sympathy, after which it returned to its former place in the flock.

A "Moderate" Drinker

The son of a Sussex shepherd, William Aylward, with regard to another notable Southdown sheep, remarked: "When my father was shepherd, a season when lambs were especially plentiful, one was given to James Pelham who kept the Oak Inn at Lavant, near Chichester. The animal was made a great pet of, being allowed to go where it pleased, and was often to be found in the cellar drinking the beer that dropped from the casks. One day Pelham, by way of experiment, turned on some extra strong beer and called the lamb to the cellar. It took a little and appeared to like it immensely, but knew when it had enough, and nothing would induce it to take any more. Upon this Pelham exclaimed, 'You be more sensible than many a human.' When about three years old the lamb had to be got rid of, as it grew troublesome (as is often the result of undue petting) and used to butt the customers."

Aylward has a store of anecdotes and interesting details respecting sheep. He tells me that Southdowns are as a rule kept within hurdles and other low boundaries without difficulty. But now and again a "jumper" is to be found among them. His father had a way of preventing such a sheep from getting out of bounds, thus :—The ear-mark is often a hole punched in the ear ; through this he passed a strong thread and sewed the ear into position to the wool of the neck. When a sheep jumps he puts his ears forward, and the jumper, on finding that his ears were fixed, relinquished the attempt.

A SHEEP MILITANT

By E. B. H., 1905

In December 1900 the headquarters and wing of the 2nd Battalion Durham Light Infantry were changing stations from Mandalay, Burma, to Wellington, Madras. They were at sea on the R.I.M.S. *Canning*, and Christmas Day was spent on board. A day or two previously sports had been held, when the officers of the ship gave a sheep as special prize for the tug-of-war. The poor fellow was destined for the Christmas dinner of the E Company, who won the event. They, however, decided to make a pet of him, and kept him as such to accompany them to Wellington, Calicut, where he remained two years, and went to England with the battalion. When the *Assaye* arrived at Southampton, "Billy" (as he was generally known, though officially styled Robert Canning, and assigned the regimental number of 9999 by his "comrades") was not allowed by the authorities to land with the battalion, so he remained in the horse-box in which he had travelled home, and was most kindly treated by the officers of the ship. Billy went to Bombay and back in this troopship

Rarer Phases of Sheep Culture

no less than three times, and was eventually allowed to rejoin his regiment at Aldershot. He became a familiar object in the 3rd Brigade of the 2nd Division, and was a most enthusiastic "soldier," accompanying the battalion on every occasion when allowed to do so. Many stories are told of his sagacity. He was never tied up in barracks, but free to go where he pleased. When marching with troops he always insisted on placing himself at their head, about five yards in front of the leading section of fours. I do not remember that he ever attempted to go to church parade; probably when he saw the red tunics he realised that it was Sunday. He used to eat almost anything, and enjoyed bananas, rice, bread, hay, corn, and other luxuries; but paper was a favourite diet of his. Alas! poor Billy died at Aldershot in January 1905, after eating a quantity of gilt and coloured papers that had been used for the Christmas decorations.

THE SHEPHERD AND HIS DOG

"MOOTIE," A SHETLAND COLLIE
(Second Prize Winner, 1909)
Property of Mr. A. J. Jamieson of Scalloway, Shetland

THE SHEPHERD AND HIS DOG

SHEEP-DOGS, PAST AND PRESENT

By WALTER BAXENDALE, 1909

THE shepherd's dog in one form or another is to be found in every country where sheep, goats, or even cattle are grazed, and the history of various varieties as they are now known is lost in obscurity. Buffon, who wrote with such authority on all pertaining to the friend of man, was of opinion that the original dog was a sheep-dog, "an animal sagacious enough to assist the shepherd to watch his flocks and herds, strong enough to protect them from ravenous animals, and ferocious enough to keep the thief and robber at a distance." The enormous dogs of Thibet and the Pyrenean sheep-dog (the biggest non-sporting dogs known) have little in common with the collie, rough or smooth-coated, and the old English sheep-dog, so well known in this country, though there is not much doubt about all being used for the same purpose ; while the herd dogs of the Himalayas and other strong, ferocious animals may be classified among the dogs which act as guardians of their masters' flocks and herds. It is rather strange, by the way, that only once in the Bible, and that in the Old Testament, is any mention made of shepherds' dogs of any kind; but in the 30th chapter of Job, verse 1, a reference is made to such dogs as follows : "whose fathers I would have disdained to have set with the dogs of my flocks." Shakespeare, however, makes no allusion at all to shepherds' dogs, though they must have been used in his time, for in pictures of the period are represented dogs bearing some resemblance to one or other of the varieties still used by shepherds and drovers. That all the varieties of the collie, including the bearded dog, the rough and smooth coated, the merle, and even the latest breed to be recognised by the Kennel Club and granted separate registration, the Shetland sheep-dog, are descended from the same stock cannot be doubted ; and one would have liked to have seen the class for sheep-dogs provided at Birmingham in 1860, the first time the breed was recognised as a show variety. All strains competed together, and one result was that from that day to this Birmingham has been looked on as a sheep-dog stronghold. One can always depend on seeing the best specimens of the variety shown there at the National Exhibition still held every winter in the hardware capital. A "pure Scotch bitch," whatever that may be, was awarded the leading prize at that first show, but the bulk of the entries were described as English sheep-dogs. There was no snapshot photography in those days, or one might compare old champions with those of the present day ; but it is certain that while the show collie and old English have been made more handsome, the best workers and the most clever dogs are little nondescript collies or shaggy bobtailed sheep-dogs that one sees at the working trials, or at such a fair as that at Findon on the South Downs, where the whole of the season's lambs are driven from outlying farms to be sold by auction. Shepherds would be badly off but for their hardy canine companions, who know what is expected of them without being reminded of their duty. To see a lamb more bold than the remainder of the flock, going too far to the right or left in search of pasture, is quite sufficient hint to them that they must be up and doing. The wanderer must be brought back to the flock, while if a little hurry is necessary when the sheep are being driven through a gate, it is the dog that is called on to impress on the stupid sheep that time must be made up by the flock being pressed and sent through the gateway a little quicker. There are several clubs formed for the purpose of encouraging the breeding of collies and old English sheep-dogs on the right lines, and classes are provided for the various strains at all representative shows though, broadly speaking, only the following are recognised : *collies*, rough and smooth coated, bearded, and merle ; *sheep-dogs*, old English and Shetland. The last-named is a diminutive collie, used in the Shetland Islands ; but the best of both English and Scottish breeders refuse to recognise the variety as a collie, and not until 1909 did the Kennel Club grant separate registration.[1]

THE OLD ENGLISH OR SUSSEX SHEEP-DOG

By The Author

"That amazing creature, the English sheep-dog."

The Southdown shepherds seem to prefer these dogs to the collie. They are "not so highly intelligent, but feel the heat less," and are less "meek." A Portland landowner describes his sheep-dogs as "shaggy, with a great deal of woolly hair over the face and eyes, as in the face of a poodle ; bluish-grey and white in colour, with

---

[1] Michael Drayton, in *The Shepherd's Sirena*, has :

"And we here have got us dogs,
Best of all the Western breed."

It would be interesting to know to what breeds Drayton and Bingley (see next page) allude.—[*Author's Note.*]

stumpy tails, and very intelligent." These dogs are noted for showing a devotion to their master and his family, to the exclusion of all others, as Mr. W. H. Hudson has recorded. And William Aylward tells me that when his father was over seventy years of age and shepherding at Lavant, he had to bring some sheep into Chichester. They got very out of hand, the old man was tired, and his Sussex dog, being muzzled, "got fogged," and refused to help him. Although the muzzling order was in full force, he was in such straits that, fearing that he would

"BOB," A SUSSEX SHEEP-DOG

"Bob" belonged to Shepherd Dick Flint on Mr. Brown's farm, Blatchington, Sussex

lose some of his sheep, he took the muzzle off, upon which the dog at once gathered the flock together, and all went well until a policeman appeared on the scene. The shepherd was fined 11s. William Bingley, writing in the year 1816, describes "the shepherd's dog" as an animal of rude and inelegant appearance, which "has its ears erect, and the tail covered beneath with long hair."[1] A friend writing from Scotland says: "You have a beautiful old breed of sheep-dog in Sussex. I have seen them in the north, where they are always called 'Sussex sheep-dogs.'"

[1] See footnote on preceding page.

William Aylward tells me the following about his father's dog: "Shepherd Aylward was very proud of his old Sussex sheep-dog, and there was great rivalry between him and another shepherd, who declared that none could beat his own dog in shepherding.

"It was 'sheep-washing' at Lavant. The sheep refused to go over the bridge across the stream which led to the pen. At first both dogs failed to manage them, for they evaded the bridge in every possible way. 'Ah!' said the Scotsman, 'your dog is not so clever after all.' But, Aylward said, 'I have not told Nimble to mount up yet.' Nimble had this order given him, and at once jumped on to the backs of the flock, running over the compact mass of woollens and snapping at their ears. In five minutes the three hundred sheep were over the bridge and penned ready for their wash. The Scotsman had to acknowledge his dog beaten. He was more accustomed to the use of a collie, which dogs do not manage their shepherding in this fashion, so he had not trained his dog to 'mount up.'"

THE MEANING OF "COLLIE"

By The Author

It is a matter of much dispute whether "collie" means "black dog" or a dog who tends the "collies" or black-faced sheep. The *New English Dictionary* gives collie as of doubtful meaning. But the Scottish use of collie for a blackface sheep is proved by a quotation in this very *Dictionary*, where we find, 1793, *Complete Farmer* (ed. 4), *s.v.* "Colley sheep," the explanation: "such sheep as have black faces and legs." To give yet other examples, *The Dialect Dictionary* says that "colley sheep," though not now used for a black sheep, was so used in the eighteenth century, and adds that "colley fleece" is the regular expression for the wool of a black sheep. Lastly, the modern usage in some parts of Scotland shows the word as applied to the sheep to be still very much alive, as the next extract will prove. The word itself means *black*, for in 1609, C. Butler, *Fem. Mon.* (1634), 122, we read of "the great Titmouse, which of his 'colly' head and breast some call a cole-mouse." In all such cases "colly" or "collie" is certainly the same as "coally" or "coaly," *i.e.* black. On the contrary, in Brocket's *Glossary of New English Words* (1825), we have "coaly, coley, a cur dog," and we find that shepherd dogs in the north of England are called "coally dogs." Everywhere in Great Britain, according to the *D. Dictionary*, the word "colley" or "coly" is used for soot, smut, coal-dust, for the black-plumaged water-ouzel, and for the old black lamp, in all of which the name is obviously given from the black colour.

Chaucer's expression, "Ran Colle our dog," has also, I believe, never been satisfactorily cleared up.

THE COLLIE-DOG OF THE HIGHLANDS

By K. Henry-Anderson, 1909

Collie dogs take their name from the blackface sheep.[1] It is not generally known that these sheep were originally called "collies" or "colleys." I was talking to a shepherd, Hafish Macpherson of Loch Shiel, Ross-shire, about this a few days ago, and he said, "Aye, that's recht, *the sheep were the collies.*" Macpherson's forebears have been shepherds in these parts for four hundred years, so that he is an authority in the matter. He seemed to be quite surprised that I should ask about it, or that any one should question the fact. *Three other shepherds* tell me that this is correct. The blackface sheep are the collies, and the dogs that tend them are in full called collie dogs, but the name of the dog has been abbreviated to collie. Quite recently in Inverness-shire I heard a girl on a farm calling to a dog "Collie dog, collie dog."

[1] This statement is, of course, based on the evidence which here follows, collected by Mrs. K. Henry-Anderson herself.

The Shepherd and His Dog

I saw some Shetland sheep in Kingussie the other day, of a delicate light brown, with cigar-coloured legs and faces—lovely little creatures. There, too, was a Shetland collie dog, fifteen inches high, light brown, with cream under-hair, which produced a kind of brindled effect; he was such a small dog, and so beautiful. Then there is also another very tiny dog, coloured black and tan, smaller than a Pomeranian. A shepherd, passing me when the tiny little "beast" had come for a walk with me, stopped and said: "Begging your pardon, mem, but isna that an awfa' wee beast o' a collie? I'm thinking it wadna be muckle use for a flock o' sheep." All I could say was that perhaps as the sheep too were so small, Barney (whose intelligence is out of all proportion to his size) might be able to "work" a little.

Beast is commonly used in Scotland for almost all animals, and frequently as a term of endearment; hence such phrases as "wee beast," "bonnie beast," or "puir beastie." It seems to have come down from the Stuart period, as *bestiole* is French for all kinds of wild creatures, birds, etc., and in Scotland "bestial" is still used in a collective sense.

THE COLLIE IN THE SOUTH OF SCOTLAND

By Sir ARCHIBALD GEIKIE, LL.D., D.C.L., 1904

The shepherds in the pastoral uplands of the south of Scotland are a strong, active, and intelligent race. I have spent many a happy day among them, living in their little shielings on the friendliest footing with them, their families, and their dogs. The household at Talla Linn Foot in Peeblesshire was a typical sample of one of these families. Wattie Dalgleish, the shepherd there when first I went into the district, was becoming an elderly man, no longer able for the stiff climbs and long walks that were needed to look after the whole of his wide charge. His young and vigorous son was able to relieve him of the more distant ground, which was shared with another man, not of the family, who slept in one of the outhouses. Wattie's active wife and daughter looked well after the domestic concerns of the household. His laugh had the clear hearty ring of a frank, honest, and kindly nature. He delighted to recount his experiences of field and fell, and his Doric was pure and racy.

Walter Dalgleish had a collie which, like himself, was getting somewhat aged, and no longer fit for the severer work of the hills. The dog would accompany him in the short rounds, and return early in the afternoon to the cottage. Some hours later I would come back from my rambles, and as I descended the steep slope opposite and came within old Tweed's sight and hearing, he would signify his recognition of me by a loud barking, which I could always distinguish from other canine performances, for it showed neither surprise nor anger, but had an element of kindly welcome in it. As I drew nearer, the barking underwent a curious change into an intermittent howl of delight, and as I came up to the enclosure the dear old creature would burst into a sort of loud guffaw. He was the only dog I ever knew that had what one might fairly call a true, honest laugh. And how his tail would wag, as if it would surely be twisted off, while he marched in front of me to announce in his own way that the guest of the family had come back. There were so many dogs in the household that one could study the idiosyncrasies of canine nature on a basis of some breadth. It struck me that perhaps there might be more truth than one had been inclined to suppose in Butler's facetious remark :

> 'Tis some philosophers
> Have well observ'd, beasts that converse
> With man take after him.

Certainly there did appear to be in that shepherd's shieling a curious similarity of disposition between the dogs and their respective masters. My old friend Tweed was a kind of four-footed duplicate of the honest Wattie, even down to the hearty laugh. On the other hand, the stranger shepherd had a collie that closely reproduced his own characteristics. The man was sullen and taciturn, did not mingle with the family but sat apart, and retired soon to his own quarter. The dog usually lay below his master's chair, refused to fraternise with other dogs, receiving them with a snarl or growl when they came too near, and marching off with the shepherd when he retired for the night. I tried hard to be on cordial terms with the man and the dog, but was equally unsuccessful in both directions.

THE POWERS OF THE COLLIE

By CHARLES ST. JOHN, 1846

The shepherds' dogs in the mountainous districts often show the most wonderful instinct in assisting their masters, who without their aid would have but little command over a large flock of wild blackfaced sheep. It is a most interesting sight to see a clever dog turn a large flock of these sheep in whichever direction his master wishes, taking advantage of the ground, and making a wide sweep to get beyond them, and then rushing barking from flank to flank of the flock, and bringing them all up in close array to the desired spot. When, too, the shepherd wishes to catch a particular sheep out of the flock, I have seen him point it out to the dog, who would instantly distinguish it from the rest and follow it up till he caught it. Often I have seen the sheep rush into the middle of the flock, but the dog, though he must necessarily have lost sight of it amongst the rest, would immediately single it out again, and never leave the pursuit till he had the sheep prostrate but unhurt under his feet. I have been with a shepherd when he has consigned a certain part of his flock to a dog to be driven home, the man accompanying me farther on to the hill. On our return we invariably found that he had either given up his charge to the shepherd's wife or some other responsible person, or had driven them unassisted into the fold, lying down himself at the narrow entrance to keep them from getting out till his master came home. At other times I have seen a dog keeping watch on the hill on a flock of sheep, allowing them to feed all day, but always keeping sight of them, and bringing them home at a proper hour in the evening. In fact, it is difficult to say what a shepherd's dog would not do to assist his master, who would be quite helpless without him in a Highland district.

Generally speaking, these Highland sheep-dogs do not show much aptness in learning to do anything not connected in some way or other with sheep or cattle. They seem to have been brought into the world for this express purpose, and for no other. They watch their master's small crop of oats or potatos with great fidelity and keenness, keeping off all intruders in the shape of sheep, cattle, or horses. A shepherd once, to prove the quickness of his dog, who was lying before the fire in the house where we were talking, said to me in the middle of a sentence concerning something else : "I'm thinking, sir, the cow is in the potatos." Though he purposely laid no stress on these words, and said them in a quiet unconcerned tone of voice, the dog, who appeared to be asleep, immediately jumped up, and leaping though the open window, scrambled up the turf roof of the house, from which he could see the potato-field. He then (not seeing the cow there) ran and looked into the byre where she was, and finding that all was right, came back to the house. After a short time the shepherd said the same words again, and the dog repeated his look-out ; but on the false alarm being a third time given, the dog got up, and wagging his tail, looked his master in the face with so comical an expression of interrogation that we could not help laughing aloud at him, on which, with a slight growl, he laid himself down in his warm corner with an offended air, and as if determined not to be made a fool of again.

HOW MASTER AND DOG CO-OPERATE

By Ralph Fleesh, 1910

There are some great, lonely characters who, by their very eccentricities, attract and amuse the public; but most biographers find it impossible to rest an enduring literary monument on a single life—they must have the broader base of associated friendships. Of the shepherd, above all others, this law of experience holds true, for any effort, no matter how sympathetic and accomplished the artist, to bring him without his collie into the limits of heroism must prove vain and disappointing. Somehow, the two have become a unit, and, as such, are charged with deep, human, nay, romantic interests. When separated their power and charm wane, and ultimately disappear.

How the two co-operate and really become one is seen and appreciated when a whole "hirsel"[1] has to be gathered from mountain and glen for shearing or other purposes. Were twenty picked athletes sent out in the early morning to accomplish the work of collecting and "bughting," the chances are they would not reach the fold till the shades of night were falling, and they might fail altogether. The sheep would play all manner of pranks with them, for, though a timid, guileless creature, the fleecy quadruped can, when opportunity offers, show a strategical resource that is simply wonderful. Sheep soon learn how to outwit man — they seldom challenge the prowess of a thoroughly trained collie. It thus becomes clear what Hogg, for instance, points out, that but for the collie sheep-farming would be an almost impossible industry. And all the wages the faithful fellow asks are three home-grown meals a day, a straw bed, and a little kindness! All great dogs, like all great men, work not because they have to, but because they want to. Action is their chief medium of happiness.

There are times when the shepherd is wholly dependent upon the saving instincts of his dog. When the snow

[1] A flock or "lot" of sheep.

has fallen quietly and heavily, and at midnight the wind caverns are opened and the fierce battalions of drift tear down the glens and up the hill-sides, like a foe that asks and gives no quarter, the shepherd must needs buckle on his plaided armour and take the field. Once out, and in the midst of the storm, the shepherd can hear or see nothing save the dull booming of muffled agony that rises from the troubled bosom of the night. He now leans upon the proved sagacity of his dogs.

Having reached the place of shelter where "drifting up" is less to be feared, and where, consequently, he wants to locate his flock, he speaks to his most experienced collie, who at once goes off at great speed in quest of his charge. The shepherd waits patiently, for complete confidence in the fidelity of his canine companion is one of the strong traits of his character. The whole world of men may deceive him—his dog, never.

At last he hears something like the low, soft sound of a waterfall, at which his young dogs drop at his feet, then rise abruptly like bundles of latent nerve touched with the soul of energy.

"Come away, man," the shepherd whispers; and instantly there emerges from the stifling gloom the old gallant with his flock.

Another sphere in which the collie figures prominently is during the lonely drives across trackless moors to the market centres, a week sometimes being spent on the journey. At night he holds vigil with his master; and should the latter seek an hour's repose, his anxious colleague will continue to move from point to point, with all the caution and care of a sleepless sentinel. When the morning breaks and the plaintive bleat of the lambs mingles with the optimistic strains of the lark, the whaup [1] in the distance trolling his own peculiar lay, master and servant meet, and on open heath breakfast together, after which they rise and move on towards their appointed goal.

The collie does not know the meaning of fear or

---

[1] The larger curlew.—[*Author's Note.*]

hardship. I was witness of the following :—A blind sheep fell into a loch and swam out a considerable distance. Fortunately, the shepherd was just returning from the "hill," and having been apprised of the accident he hurried with his three collies to the rescue. He simply whispered a syllable to one of his dogs, who at once took to the water. Out she went—tried to bring the sheep ashore one way, then tried it another. The shepherd stood motionless; his two young dogs impatiently whined at his feet. At last somebody cried, "Auld Rasp" (meaning the dog) "is gaun tae be drooned." "Yes," replied the stern-featured, stalwart mountaineer, "she will die or save her charge." After the finest display of sagacity under most trying circumstances I ever beheld, she brought the sheep to the bank, she herself being so exhausted that her master had to lift her out of the water, carry her home, and administer restoratives. "Rasp," the heroine of this scene, belonged to Mr. James Gardner, of North Cobbinshaw, Midlothian, and deservedly ranks as one of the great dogs of history.

Many stories are told of "Old Rasp," for to her memory all who knew her are ever ready to pay tribute. On one occasion a pig, which had been brought home the previous day, escaped. The sun was setting when Mr. Gardner returned from the moors. Finding "the guidwife" much excited over the abrupt departure of the little stranger, he allayed her fears by assuring her that "Auld Rasp will soon bring back the wee prodigal." So off Rasp went in quest of what proved one of the most stubborn of the members of the bucolic family she ever encountered. Having been absent about twenty-five minutes she at last appeared with a few sheep in front of her. But in the centre of the sheep was the pig, experience having taught her that the little rebel could not be driven alone. Ever afterwards she visited the sty daily to make sure that the occupant was being kept in his proper sphere.

On one occasion Mr. James Gardner was taken ill while bringing home a flock of sheep. A friend having secured the sheep in a park had him driven home; but "Wharry," another of his famous dogs, was kept that she might assist in conducting the sheep home the following day, a distance of between five and six miles. It was a wild stormy night in the month of February. "Wharry" was so anxious about the safety and progress of her charge that she made a determined effort to get free, and succeeded. About midnight she was found with the sheep all gathered up to the gate of the park, and there she lay patiently waiting for her master. The following morning when the gate was opened she started for home with the flock, like something endowed with human intelligence. When she had arrived, and had received the greetings and "Well done" from her master, she lay down on the hearth, and seemed the most pleased and satisfied of toilers.

To the sagacity of the thoroughly trained collie there is indeed no limit. It has been our privilege to be closely associated with the greatest sheep-dog trainer Scotland ever produced, and we have heard him repeatedly say of his favourite dogs that their intelligence was always more than equal to any emergency. "When riding in South America," says Darwin, "it is a common thing to meet a large flock of sheep guarded by one or two dogs, at a distance of some miles from any man or house." This is not at all extraordinary. We know a dog, the property of a shepherd already referred to, which took charge every morning of a certain "cut" of sheep and had them directed through gates and over hedges to a lowland pasture some three miles away. He needed no bidding or exhorting; he had learned the art of dignifying service.

Our "born" shepherds—the true sons of the calling—do not forget their old canine colleagues. Travelling in the sheep districts of Scotland, an old corpulent collie, long retired from the stern duties of the "hill," lying on the green sward in front of the shepherd's cot, is quite a common sight. If the day is warm you may find the shepherd's child sleeping in his bosom. The mother has no hesitation in leaving the infant so watched and protected; for the old retainer, having been the first object of the child's curiosity and love, gallantly responds with

Photograph by A. Brown & Co., Lanark.
"FRISK" (A SHORT-HAIRED "BEARDIE"), "THE CHAMPION OF THE WEST," AND MR. ALEXANDER MILLAR OF BURNFOOT, AYRSHIRE (OWNER)

an instinctive gratitude by assuming responsibility for the safety of his youthful charge when the pressure of circumstance demands. And when the old and faithful friend comes to die, deep and sincere is the lamentation of the whole family. We have seen a shepherd with the dauntless courage of a lion kneel by the side of his dead companion, and bewail his loss like a grief-stricken boy.

HOGG'S "FAITHFUL SIRRAH AND HECTOR"

By James Hogg ("The Ettrick Shepherd"), 1772-1835

About seven hundred lambs, which were once under my care at weaning time, broke up at midnight and scampered off in three divisions across the hills in spite of all that I and an assistant lad could do to keep them together. "Sirrah, my man," said I, in great affliction, "they are awa." The night was so dark that I could not see Sirrah, but the faithful animal heard my words—words such as of all others were sure to set him most on the alert—and without much ado he silently set off in search of the recreant flock.

Meanwhile I and my companion did not fail to do all in our power to recover our lost charge. We spent the whole night in scouring the hills for miles around, but of neither the lambs nor Sirrah could we obtain the slightest trace. It was the most extraordinary circumstance that had occurred in my pastoral life. We had nothing for it (day having dawned) but to return to our master, and inform him that we had lost his whole flock of lambs, and knew not what had become of them. On our way home, however, we discovered a body of lambs, at the bottom of a deep ravine called Flesh Cleuch, and the indefatigable Sirrah standing in front of them, looking all around for some relief, but still true to his charge. The sun was then up, and when we first came in view of them we concluded that it was one of the divisions which Sirrah had been unable to manage until he came to that commanding situation; but what was our astonishment when we discovered by degrees that not one lamb of the whole flock was wanting! How he had got all the divisions collected

The Shepherd and His Dog

in the dark is beyond my comprehension. The charge was left entirely to himself from midnight until the rising of the sun ; and if all the shepherds in the forest had been there to have assisted him they could not have effected it with greater propriety. All that I can further say is, that I never felt so grateful to any creature below the sun as I did to my honest Sirrah that morning.

HECTOR, THE SON OF SIRRAH

He was the son and immediate successor of the faithful old Sirrah ; and though not nearly so valuable a dog as his father, he was a far more interesting one. He had three times more humour and whim about him ; and though exceedingly docile, his bravest acts were mostly tinctured with a grain of stupidity, which showed his reasoning faculty to be laughably obtuse.

I shall mention a striking instance of it. I was once at the farm of Shorthope, on Ettrick Head, receiving some lambs that I had bought and was going to take to market with some more the next day. Owing to some accidental delay I did not get final delivery of the lambs until it was growing late ; and being obliged to be at my own house that night, I was not a little dismayed lest I should scatter and lose my lambs if darkness overtook me. Darkness did overtake me by the time I got half-way, and no ordinary darkness for an August evening. The lambs having been weaned that day, and being of the wild black-faced breed, became exceedingly unruly, and for a long while I lost hopes of mastering them. Hector managed the point, and we got them safely home ; but both he and his master were alike forfouchten. It had become so dark that we were obliged to fold them with candles ; and, after closing them safely up, I went home with my father and the rest to supper. When Hector's supper was set down, behold he was awanting ! and as I knew we had him at the fold, which was within call of the house, I went out and called and whistled on him for a good while, but he did not make his appearance. I was distressed about this ; for, having to take away the lambs next morning, I knew I could not drive them a mile without my dog if it had been to save the whole drove.

The next morning, as soon as it was day, I arose and inquired if Hector had come home. No ; he had not been seen. I knew not what to do ; but my father proposed that we should take out the lambs and herd them, and let them get some meat to fit them for the road, and that I should ride with all speed to Shorthope to see if my dog had gone back there. Accordingly we went together to the fold to turn out the lambs, and there was poor Hector, sitting trembling in the very middle of the fold door, on the inside of the flake that closed it, with his eyes still steadfastly fixed on the lambs. He had been so hardly set with them after it grew dark, that he durst not for his life leave them, although hungry, fatigued, and cold, for the night had turned out a deluge of rain. He had never so much as lain down, for only the small spot that he sat on was dry, and there had he kept watch the whole night. Almost any other collie would have discerned that the lambs were safe enough in the fold, but honest Hector had not been able to see through this. He had even refused to take my word for it, for he would not quit his watch, though he heard me calling both at night and morning.

THE SHEEP-DOG OF IRELAND

By Ralph Fleesh, 1910

In appearance the Irish sheep-dog strongly resembles the old Scotch Border collie—the "bobtail," though seen in some parts, is not common ;—strong in instinct, and trained to perfection, this most human of all animals is capable of great feats. He will run out a mile, or even a greater distance if necessary, for a lot or "cut" of sheep; bring them cautiously up to his master's feet ; then he will assist at shedding or penning, as the case may be, with a will and wisdom that gives him a superior status in the animal

The Shepherd and His Dog

world. Shepherding in Ireland, as in other countries, without a sheep-dog would border on the impossible; like sheep-farming without the shepherd, it would be as the tragedy of *Hamlet* without the Prince.

SHEEP-DOGS (*Coill*) IN THE ISLE OF MAN

By The Author

Shepherd Caley of Ramsey tells us that the old Manx sheep-dog was a "holding," not a driving dog. It kept to heel, and when a particular sheep was wanted, the shepherd would point to it and say in Manx, "There, Spring, go and hold that rough fellow," and the dog would seize the sheep behind the neck, throw it down, and hold it with his paws, never hurting it. These dogs are now extinct in the island; they did not work the sheep as the collie does. They are described as smooth-haired, of various colours, very big and strong. Dr. Tellet of Ramsey writes: "I recollect having seen one of these dogs about sixty years ago, which belonged to an old man who lived near Ramsey. It was smooth-haired, and my impression is that it was about the size of a Scotch deerhound, coloured black, grey and tan—the tan so intimately mixed with the grey in parts as to produce a rust colour. I see the colour in the dogs we have now, a number of which are descendants of crosses between the dog in question and the Scotch collie. The collie is said to have been brought to the island by the Scotch shepherds who came over to take charge of the larger sheep farms. The first I saw on the island I remember distinctly; it was black and white with very silky hair. A Mr. Metcalf had the credit of introducing the old English sheep-dog also at this date. The collies were not generally used until about 1860. I have heard my father say that the Manx dog was only a holding dog. A few days ago I was talking to an old shepherd, who described the way it threw the sheep down." Miss Sophia Morrison of Peel writes: "Some years ago a Manx shepherd told me some wonderful tales of an old sheep-dog. This shepherd used to go to the mountains with his father to look after the sheep, and his father had only to point his finger at any one sheep in the flock and say, 'Grein yn nane shoh, Coly' ('Seize that one, Coly'), or 'Greim mee shen' ('Seize that for me'), and the dog at once put his paws on the sheep pointed out to him in the midst of the flock, and held it till the old man came up." Another person remembers sheep-dogs not in the least like the sheep-dogs of to-day ; they were larger, smooth-haired, and were known as "houl'ers, because they were good to houl' on." These dogs upset the sheep on to their backs and kept them down until the shepherd came to them. This old shepherd did not think that these were native sheep-dogs, but that they had had special training to make them "houl'ers." An authority in the island remarks "that if there had been a breed peculiar to the Isle of Man, some of the historians who wrote about the native pony, sheep, and cat would have mentioned it. They were probably introduced by the Norsemen, and existed in other places in the United Kingdom at the same time, *i.e.* about fifty or sixty years ago, and were only sheep-dogs by special training."

Ralph Fleesh tells us that the dog applies his mouth to the wool as well as his paws to the neck, but the skin of the sheep is never injured; adding, "To upset a sheep is a mistake, since the process involves a shock that sometimes leaves bad results. I knew a 'beardie' collie named 'Roy'—one of the heroes of his day—who could hold up any sheep without upsetting it. He was a powerfully built dog, and so by seizing the wool of the sheep's neck, and meeting by quick movements every effort of his charge, his strength and weight being a sufficient barrier, complete victory was easily and promptly achieved." There are local shepherds' dogs in various districts in England, Ireland, Scotland, and Wales, as well as in the Isle of Man, but these local breeds cannot be regarded as distinct, since they lack uniformity of type.

THE SHEPHERD'S AND DROVER'S DOGS COMPARED

From *Pictorial Half-Hours*, 1851

Closely allied to the shepherd's dog is the cur, or drover's dog. This useful animal is larger than the shepherd's dog, the hair is generally shorter, and the tail, even when not cut purposely, often appears as if it had been so. Bewick, who was well acquainted both with the drover's and the shepherd's dog, speaking of the former, says: "Many are whelped with short tails, which seem as if they had been cut, and these are called in the north 'self-tailed dogs.'" The same writer is disposed to consider this breed as a true or permanent kind, and he informs us that great attention is paid to it. It seems to us, however, that the drover's dog is in reality a cross between the shepherd's dog and some other race, perhaps the terrier. It often partakes largely of the character of the shepherd's dog, but is taller in the limbs. These dogs are singularly quick and prompt in their actions, and, as all who have watched them in the crowded, noisy, tumultuous assemblage of men and beasts in Smithfield must have observed, they are both courageous and intelligent. To their masters, who often ill-treat them (the drover has not always the kind heart of the shepherd), they are faithful and attached.

THE DROVER'S DOG

By Edward Jesse, 1853

There is reason to suppose that tamed dogs of whatever species which were first employed for any useful purpose were employed as sheep or shepherd's dogs, because we are taught by history to conclude that men were shepherds before they were hunters;[1] and because the great use of a dog in the operations of a shepherd is suggested by his

---

[1] The author seems to be mistaken; it could hardly be maintained seriously that the pastoral life preceded that of the hunter.—[*Author's Note.*]

sagacity and obedience and the instinctive fear of him which has been implanted in sheep, a fear that never diminishes by experience, but operates equally upon sheep of all ages.

The terror of sheep at the bark of a dog is so great that when they have learned to associate it with a loud whistle, as they do when they have travelled in a drove, they will be terrified as much by the sound of a loud whistle as by the loudest barking of a dog, and run together when they are driven by a man who whistles like a drover, in the same way as when they are collected by a dog.

I was interested the other day in watching a flock of sheep, attended by a drover and his dog, as they were passing along a turnpike road. The man went into an ale-house by the roadside, leaving his dog to look after the sheep. They spread themselves over the road and footpath, some lying down and others feeding, while the dog, faithful to his trust, watched them carefully. When any carriage passed along the road, or a person was seen on the footpath, the dog gently drove the sheep on one side to make a passage, and then resumed his station near the ale-house door.

Those indeed who have travelled much at the time of the great fair of Weyhill must have observed the sagacity of the drovers' dogs on the approach of a carriage. A passage is made for it through the most numerous flocks of sheep in the readiest and most expert manner, without any signal from the drover. The fatigue that these dogs must undergo is very great. One sees them sidle up to their master after each exertion, and look at him, as if asking for his approbation of what they had done.

When I occupied a small farm in Surrey I was in the habit of joining with a friend in the purchase of two hundred Cheviot sheep. The first year we had them the shepherd who drove them from the north was asked how he had got on. "Why, very badly," said the man; "for I had a young dog, and he did not manage well in keeping the sheep from running up lanes and out-of-the-way places." The next year we had the same number of sheep brought up, and by the same man. In answer to our question about his journey, he informed us that he had got on very well, for his dog had recollected all the turnings of the road which the sheep had passed the previous year, and had kept them straight the whole of the way.

TRAINING THE SHEEP-DOG IN ENGLAND

By H. Somerset Bullock, 1909

In the Sheep Pool, towards which the flock are threading their way, the last lingering lights of the sunset are faintly mirrored, until the clouds curtain the fading gold and pink in the western sky, and one can see the wings of darkness fold over the weald. The bleating of a sheep and the short, sharp bark of the dog—and then the night silence. It was just one such evening that I walked home with my friend the shepherd—a tongue-tied man, you would say, until you knew him. But you must remember that time ploughs slowly along for him and leaves much leisure for thought. The shepherd's life has changed less with the change of years than that of any other calling. It needs a man who has grown gentle and almost motherly to be a shepherd—there is so much mothering to be done.

Scarcely less interesting than the shepherd himself is his dog. In answer to my questions as to the training necessary, my friend told me several facts that were new to me.

"When he is a pup," he said, "we teach him to obey his master's call at once. It must be done with kindness, or the dog won't love you as he must if he is to serve you well."

"But how is it you manage to prevent his biting the sheep?"

"Well, he has to learn slowly. It's natural for him to want to gather any animals together if he sees them wandering apart, and an untrained puppy will do it of his own accord. But, as you say, he would worry the sheep.

I taught Bob by making him lie down, no matter whether he happened to be close beside me or several hundred yards off. At the word he would do it, so that I could always stop him from chivying when I liked. If he'd been very bad at it I should have tried with turkeys; they would soon teach him not to snap. If not, we have to muzzle him."

"Does he understand quickly the object of keeping the sheep in a flock?"

"Watch him now," returned the shepherd. He gave a sharp, clear whistle, and the dog, which up to that moment had been leading the sheep, immediately came to heel. The sheep hesitated, looked doubtfully ahead, and stopped short. Again a whistle, and the dog "rounded" the flock, beating up the stragglers. Yet a third whistle, and he again took the lead, the sheep following him.

"Have you taught the sheep to follow?"

"No," he laughed; "all sheep will follow their leader, even if he happen to be a dog."

"How much will you take for him?" I nodded towards Bob.

"Don't ask me, sir," he almost pleaded; "there's another chap sold his for twenty pounds to a gentleman the other day, and he's been sorry ever since. No, sir; Bob sticks to me, and though you offered me thirty pounds I'd stick to Bob."

TRAINING THE COLLIE PUP IN SCOTLAND

By Ralph Fleesh, 1910

He is reared in the kitchen of the shepherd's home. When about two months old the shepherd, after the labours of the day, takes him in hand, the meaning of language having already been taught him, as a rule, by the shepherd's children. The little fellow, though rebellious at first, soon gives fine point to the law of obedience by responding to every whisper and signal of his master. He is made to "clap down," and lie firm with his head close to the floor between his fore-paws. (Of all the attitudes of the collie when in action this is the prettiest.[1]) Then his master calls him up, makes him move to another point, and "drops" him again. When the tiny canine pupil promptly honours all the commands, he is congratulated by the whole household and made the hero of a little banquet. Once thoroughly trained in this fashion, the shepherd has little or no difficulty with him in the open. A puppy "of the right kind" clings to his first master, provided, of course, that master is worthy. The greatest dog trainer I have known had periodical visits from all his pupils after they had been returned to their owners. Ofttimes have I heard him say, when some poor fellow after a long journey would bound into the kitchen, embracing all the members of the family in turn, "Puir chap, I wish you could stay." This shepherd's favourite maxim was, "Make a dog love you, and he will never fail or forsake you."

SHEEP-DOG TRIALS IN ENGLAND, WALES, AND SCOTLAND

By Walter Baxendale, 1910

Many improvements in the management of sheep-dog trials have been introduced since the institution of these very interesting competitions by Mr. Lloyd Price of Rhiwlas, near Bala, in 1873, but the actual tests have remained almost the same; whether the meeting is at Tring in the south, or in Caithness, the highly trained sheep-dog is expected to collect his little flock, drive the sheep through various obstacles, and finally pen his charges. During the past ten years the trials have become wonderfully popular; and though Lord Leconfield did not persevere with the meeting he established in connexion with his tenants' agricultural show at Petworth in 1908, yet other southern fixtures at Tring, and also in

[1] Even when he is motionless physically, the mind of the dog is expressed by his eye in commanding sheep.

connexion with the perambulatory show of the Bedfordshire County Agricultural Society, have flourished, and the trials are certainly among the popular features of either show. For real enthusiasm, however, you must go to the north, or to one of the several good meetings held in Wales during the season, though at the last-named the southerner is puzzled to hear the shepherds yelling directions to their dogs in a mixed jargon of Welsh and English. Courses for trials vary; but the ground should not be too level—it is far better for the purpose if undulating, while the task of the dogs is made more difficult, and they are called on to exercise greater care in the driving and collecting of their flock if there is a little stream or burn to be crossed. Gaps can be made in hedges, or rows of hurdles arranged so that an opening is left for the sheep to be driven through, while at most of the first-rate meetings an obstacle known as the Maltese cross is introduced before the pen is reached. It is at this puzzle that the shepherd is generally allowed to leave the spot from which he has directed the work and assist his dog at closer quarters, for no obstacle requires more careful negotiation, and many a trial has been won or lost at the cross. Hurdles are arranged in the shape of a Maltese cross, the dog having to drive the flock through the two strands or rows, and then up to the pen, before his task is completed. The almost perfect understanding existing between man and dog is the most remarkable feature of a sheep-dog trial, and no sporting dog of any kind answers so well to the call of his handler as the little unkempt working collie. He may have no beauty to recommend him, some of the best workers being mongrels, though such men as Barcroft of Bury and Akrigg of Sedbergh, who have gained scores of honours in competitions all over the country, have succeeded in establishing a certain strain. "Handsome is as handsome does," however, and watching the work of the dogs one marvels at their intelligence and admires the patience of their handlers. The sheep may be released six hundred yards or even more away, but at the words, "Go to 'em, Jess,"

The Shepherd and His Dog   151

the dog—if well trained—makes a bee-line to where they are grazing, and, ranging almost like a setter, she gets a good distance behind her sheep before beginning to drive them. "Closer in, lass," calls the handler, or he whistles in a key which is understood by the dog to mean the same thing, and then the task begins. The sheep may be inclined to "split," but Jess is prepared for that, and

"BEN" (A ROUGH GREY MERLE) AND MR. THOMAS GILHOLM (OWNER)

"Ben was the winner in a number of 'Trials.' When eight months old he won third prize out of twenty entries, and after that was never out of the prize list, winning many firsts. For steadiness and style he could not be beaten."

reaching them she gets them together again, and on being motioned by the handler, changes her position so that the sheep can be driven once more. She seems to understand that the flock must not be too closely pressed, and one by one the obstacles are safely negotiated and the pen is reached.

This is a crucial test, and it is always amusing to notice the antics indulged in by the sheep. They walk round and round the pen as if looking for the entrance, though quite disregarding it till, frightened by the shouts of the shepherd and the knowledge that there is a dog not far away, one sheep gets her head in the pen. The shepherd becomes excited and calls for Jess to be steady; if she is one of Barcroft's famous strain she crawls on her belly and literally creeps to her sheep. By this time the ewe has made up her mind to go into the pen, and if she can be kept there the other members of the flock are sure to follow. The task is at last completed, the timekeeper notes how long Jess has taken, and reports to the judge, who, adding points for the style in trying circumstances, compares her performance with that of her closest competitors. Another dog is called for, and the trials then proceed. It may be added that at many of the best meetings the shepherd directs the early stages of the work of his dog from a circle marked out by flags, in the middle of which is a stake, and to that stake is fastened a cord which must be held by the shepherd till he is given the signal to go and assist his dog. That is generally when the Maltese cross is reached.

A LIGHT OF OTHER DAYS

By C. Brewster Macpherson, 1909

A wanderer on the Highland Border, I found myself recently within reach of an important sheep-dog trial meeting. Arrived there, I was soon deeply interested in the marvellous work of some of the best dogs from the Border. As I changed my position for a better view, my attention was caught by a solitary sheep-dog bitch, a matted tangle of an uncared-for thing. She lay apart by herself. Occasionally she rose, and pushing her way through impeding legs, gravely contemplated, with critical eye, the various operations which were being carried out in the shedding, penning, etc., but made no attempt at assistance, except when sheep which had been through the course were being taken by the last competitor to a field at the side, into which they were turned. On several of these occasions, for no especial reason that I could see,

The Shepherd and His Dog

she would join in herding them in a very business-like manner, and then return to her original position. Knowing the rule which requires all competing dogs to be on the lead, I thought she must be some drover's bitch, some hanger-on of the auction mart who had come to see if dogs were as good now as they were in her day. I forgot all about the trials, and found myself continually watching her instead. I tried to scrape an acquaintance, asked her who she was and where she was from ; but while receiving my advances with courtesy, she gave me clearly to understand she did not desire an intimate acquaintance. Thinking she looked bored, I sought a bit of meat for her in the tent. After regarding my face earnestly for some time, she took it, but her manner conveyed a world of rebuke, and feeling I had done quite the wrong thing and advanced myself no whit in her estimation, I retired abashed by her cold demeanour. The card is run through, a loud voice proclaims that an entry which had been overlooked in error will be allowed to run, and a broad-shouldered, good-natured-looking Borderer steps over the ropes—but where is his dog ?

A chirrup! the tangled mat is alive, her lethargy gone ; trembling with suppressed excitement, at his side she stands, the wisdom of ages in her beautiful eyes, which are fixed intently on her master's face. And then followed the most masterly exhibition which the writer, who has judged at many such trials, ever witnessed. And though she only obtained the second place of honour, she confided to me, in a farewell interview, that she had never worked better, and did not see how the thing could be done better, with which opinion I most cordially agreed. An attempt to buy this light of other days was quenched by, "Na, it's no that a'thegither, but ye see they a' come aifter her pups an' I mak a gey bit oot o' her yon way."

AULD KEP: "A PAST-MASTER," AND "ONE OF THE GREAT DOGS OF HISTORY"

(A Pure-Bred Border Collie)

THE PROPERTY OF MR. JAMES SCOTT

By RALPH FLEESH, 1909

Auld Kep—for this is now his familiar name—the winner for the second time of the International Cup, is an average-sized dog of the type of the old Border collie. He is finely coupled, and in action shows to great advantage at the sheep-dog trials. When he leaves his master to take command, there is an ease and confidence revealed that instantly stop the flow of speculations. The sheep seem at once to recognise his kindly powers, and, instead of rebelling, comply with his every request. Having an extremely strong eye, he at close quarters throws a mesmeric influence over both sheep and spectators. Now well accustomed to the trial course, he keeps perfectly cool, carefully scans the ground before beginning, and then lends an attentive ear to his master, and to his master only, no matter the excitement and noise beyond. When scarcely a year old he came to Mr. Scott's hands, having then a deal to learn. He has won considerably over £200 in prize money, besides cups and medals. To-day all authorities recognise him as the greatest sheep-dog living. That true working blood courses through his veins, is shown by the fact that his sons and daughters filled the entire prize list at the late sheep-dog trials at Perth on September 18. He is now eight years old.[1]

[1] For a portrait of Auld Kep, see title-page. It is inserted by the kind permission of Mr. James Scott of Ancrum, and of the Editor of *The Field*.

A SCOTTISH SHEEP-DOG TRIAL. REPORTED IN "DORIC"

A VETERAN'S VIEW

By Ralph Fleesh, 1909

Dear Mister Editor—By the mysterious decrees o' Providence it is my lot to be wedded to Peggy—a woman o' great geefts an' extraordinar' ambeeshun. Alexander the Great, Ceecero, an'—no to gang sae far back—the Airl o' Chatham, were, I am tauld, a' seemilarly circumstanced. Indeed, if I can read history richt, men who, like mysel', have become great in the world's affairs, owe a deep debt o' gratitude to their wives. We may occupy thrones, but, mind you, Mister Editor, weemin put us there. The cheery, delightfu' craeturs may hae led us a wee bit astray at first, but I dinna believe in the heathen practice o' openin' auld sairs.

Weel, Peggy said to me last Thursday, that seein' I was an authority on dowg-workin' I should gang (in her company) to the great Internaitional Sheep-dowg Trials at the Fair Ceety o' Perth on the Seturday. She made the proposal in such a coaxin' kin' o' a wey, bringin' me in min' o' ither days, that I at once agreed. "Noo," she said, "ye maun hae a suit o' claes befeettin' the occasion, an' which, at the same time, will be prophetic o' yer destined station in the world o' fashion"; so Peggy an' me plunged into kist efter kist, at last findin' what we were seekin'—a rig-oot o' velveteens adorned wi' blue braid, the purchase for an auspeeshus occasion fifty-five 'ear syne. Peggy resolved to put on her blue Frainsh mareena, in order to effect a proper an' becomin' match.

Off we went on Seturday mornin', Peggy steepilatin' that we would breakfast in an hotel at Perth, jist for the look o' the thing, an' that we micht be able to tell oor neebors when we cam' hame a' aboot the denties and ongauns o' high life. A' the wey north, herds and farmers cam' poorin' in at ilka station, until Peggy declared she was convinced the train would stick. But no; it puffed and bowff'd away in a praiseworthy, though no very musical mainner, arrivin' forty meenits late, in a highly exhausted condeeshun.

Havin' had breakfast—an', my word, Peggy made the waiters staun' aboot—we took a brake oot to the trial field. The man wha drove us was very polite, addressin' Peggy as madam, an' me as sir, the result being a monetary loss to me; for, says Peggy, efter he had lifted her off in his airms, at the entrance gate, gi'e the puir chap a sixpence extra to buy sweeties for the bairns, for, by the complexion o' his coat, I guess he has a fine swarm o' them. I promptly obeyed.

When we got inside we were fairly amazed. The crood was tremendous, everybody seemin' mair happy an' excited than anither. Motor car efter motor car cam' rollin' in until I commenced to wonder whether there had ever been such a graun' an' glorious show afore. A tup sale, honestly, Mister Editor, is a puir thing in compairison.

Makin' oor first roon', wha did we meet but Jamie Scott wi' Auld Kep, the greatest dowg leevin' (accordin' to "Ralph Fleesh"), an' twae o' Auld Kep's progeny. Peggy was awfu' pleased to meet Jamie an' hae a look at the Internaitional champion. "Mexty me, Samil," she exclaimed, "he's jist the very brither" (referrin' to Kep) "o' oor auld Toss, an' I'se warrant he's nae better." At this Jamie smiled sae sweetly that Peggy declared efter we left that she wasna surprised in the least at the Heelant lassies fa'in' in love wi' him.

Rememberin' that Peggy had seen mair birthdays than she cared to admit, an' that she wud be the better o' a sait, I conducted her past the ropes, straucht into the preevileged enclosure where the cream o' the nobeelity were sittin'. Although a wee bit excited, we mairched boldly alang, keepin' oor een open for vacant places. There no' bein' a spare inch, we were jist aboot to retire, sairly disappointed, when a man wi' a white waistcoat rose —his size needin' accommodation for twae—took off his hat—nae doot as a mark o' respect, but, at the same time, to show off his fine heid o' hair—an', says he, in a maist

MR. W. B. GARDNER, SHEEP-DOG JUDGE

"That's 'Ralph Fleesh.'" "Preserve me!" she exclaimed, "is that the man? What a lot it maun hae ta'en to bring him up!"

eloquent fashion, jist as if Peggy an' me were a public audience, "Allow me, lady and gentleman, to accommodate you." We at yince responded, I raisin' my hat, though no wi' near as fine a flourish as the exheebitor o' the white waistcoat. Yince firmly fixed, an' wonnerfu' comfortable, I says to Peggy, "Dae ye ken wha that obleegin' chiel is?" "No," she said; "but he micht be onything frae a theatrical to a horse-cowper." "Weel," I replied, "that's 'Ralph Fleesh.'" "Preserve me!" she exclaimed, "is that the man? What a lot it maun hae ta'en to bring him up!"

But "Ralph," who is not contemptuous o' tactics, put himsel' to no end o' trouble to win the favours o' Peggy. He brocht up the chairman an' the secretary o' the Internaitional Trials Society—twae as fine-lookin' men as I ever clapped een on—an' introduced them to Peggy an' me. Then he went an' got the Coorse Director, the man wha kens a' thing—a stern, meelitary-lookin' gentleman wi' a regilar field-mairshal voice—an' Tam Gilholm—a perfect jewel o' a man, wi' a parteeklar nice hame-spun suit o' claes—an' likewise introduced them. Needless to say, this attention greatly pleased Peggy; nor did I, Mister Editor—if the truth must be spoken—feel bored wi' his civeelities.

Three solemn-lookin' mortals sat on a sait in front o' us, whose awsome appearance at yince aroused the curiosity o' Peggy; for, "Noo, wha can they be, an' what crime will they hae committed?" she queried. "Wheesht, lassie," says I; "thae men hae the poo'er o' kings—tha'se the jidges!" "Help us!" retorted Peggy; "they're mair like culprits waitin' the approach o' the hangman." Clever, far-seein' woman though Peggy is, she failed to grasp the full signeeficance o' their heavy responsibeelities. Had she less o' a masterfu' mind hersel'—for hers is capable o' governin' nations, let alane me—she would appreciate the leemits an' deefficulties o' ordinary mortals wi' a truer sense an' a keener sympathy. Cromwell could never, for instance, be expected to understaun' the worries an' leemits o' his valet. The same wi' Peggy an' dowg-trial jidges—she's sae far abune them, intellectually, that what should be peety, on her pairt is apt to become scorn.

The Shepherd and His Dog

The trials were noo in full swing, yae herd efter anither tryin' his luck—for there is a guid deal o' luck in't—wi' his speerited an' faithfu' companion. Much, of coorse, depends upon the skill o' the dowg, but .the temper o' the sheep is also an important factor. Peggy quickly noted this, an', as was her richt, went ower an' told the jidges their plain an' bounden duty in the maitter. They a' raised their hingin' heids, like men wha had been dookin' for aiples, an' looked queer at Peggy, which made her speer whether they were sufferin' frae heidaches. "Oh no," the spokesman said; "but if ye staun' lang there we'll be shair to experience that affliction"; whereupon Peggy withdrew, for, as she said, it was quite obveeus her presence was extremely tryin' to their nervous system.

The fine workin' o' the dowgs I'll no try to describe, for, railly, there were some rins sae bewitchin' an' perfect that my pen wud falter far ahint the reality. The hale affair was a noble an' matchless pictir' o' sublime, etherial motion. Peggy, wha is awfu' poetic at times, said it was sae sweet that she was frequently lulled into dreamland; but aye when her een were jist aboot to close, a shout frae the chap wi' the field-mairshal voice brocht her back to her senses. "Samil," she said, efter an unusual roar that cowpit an auld man aff an end sait, "if you an' that chiel were Jainerals in the Territorials the Germans micht weel trimil."

Although the trials lasted the hale day we didna weary, for there wasna a dull meenit. When the results were announced there was deefinin' applause, for everybody seemed delighted that Auld Kep had again carried off the cup. But there were twae omissions in the prize list that painfully surprised Peggy an' me. Hoo Tammy Broon, Tollishill, wi' Lad, an' Tammy Armstrong, Pinnacle, wi' Moss, were keeped oot fairly passed oor comprehension. Peggy was of the opeenion, an' is of the opeenion still, that the jidges, fateegid an' worn oot wi' the labours o' the day, fell soun' asleep when they were rinnin'. She says she actually heard them snore, which,

I think, must be a wrang impression, for the field-mairshal chiel was never far frae their lug.

Whatever the explanation, Mister Editor, Peggy an' me were very very sorry for Lad an' Moss, seein' they had worked sae cleverly. But jidgin' dowgs is jist like jidgin' sheep—it's a kin' o' a lottery, aboot which the less said the better.

Peggy, however, was sae vexed aboot the deceeshun that she threatened to become obstreeperous; but, thanks to Mr. Clark, the chairman, his fine speech at the close brocht her back to a calm an' reasonable frame o' mind. Seizin' the opportunity—for I'm a great domestic strategist—I took Peggy's airm an' said in the auld sweet wey, "Come awa' hame, my darlin'," an' sae, Mr. Editor, we left.—Believe me, yours in hame-spun sincerity,

SAMUEL WHEEPLETON.

SHELTER-NEUK, PLAIDYPLINKS,
27th September 1909.

"MAGNUS" AND "RONALD"

By MAX PHILPOT, 1909

" Heard ye the news, Davie ? "

" What news ? "

" Aboot the Internaitional next week."

" Weel, what new terror is threatened us noo ? "

" The great Suffolk dowg's booked; an' so is the Irish champion; an' I had an inklin' frae Maister Whinny, the secretary, the other day, that the dowg that carried off the £100 and Colonial Cup in New Zealand twa month syne is on the water bound for the Internaitional at Perth on Wednesday."

" Indeed, Samil, indeed ! Nae doot it 'll be a stiff struggle, but I'm no to tak fricht at the tootlin' o' far trumpets. Auld Ventir (Venture) has met the best o' his time, an', puir auld chap, though, like masel', no sae glossy and swak as he yince was, he'll lower the tails o' a few o' them yet."

The Shepherd and His Dog

So talked Samuel Tweedie and David Garlow, two well-known Pentland shepherds, the latter being the trainer and master of Venture, the famous collie whose premier record at National and International Trials had never been broken.

While the two mountaineers thus sat and conversed, Venture lay asleep on the hearth, now and again giving troubled expression to the reflex action of spent emotions, Mary (Mrs. Garlow) taking great care, during her preparation of tea, not to disturb the old giant's repose.

"He's dreamin'," said Davie; "for after a hard day's work the peace o' the body is seldom followed by a perfect peace o' the mind."

That David Garlow had perfect confidence in the prowess of Venture there could be no ground for doubt; but that he anticipated a more than ordinary challenge at the great International Trials the following Wednesday was also beyond dispute. As the day drew near a few of Davie's most intimate friends observed symptoms of anxiety in his manner, the utter absence of which had long been the most noted feature of his character. Doubtless, the threat by Mr. Blacklock Garston, the well-known dog fancier (who, by a vote of the members of the International Society, had been removed from the Executive Committee, because of his having dared to criticise publicly the awards, and the nature of the work and course at the previous trials), that he would lower the pride of the south-country men by producing a dog which would set a higher standard of work than had yet been seen—this, together with sundry premonitions, had opened up to Davie's vision wide fields of possibility in which he sometimes saw the threat of eclipse.

"But," he would say to himself, "this is a' idle imagination on my part, as I'm convinced it was only brag and bluster on the part o' Maister Garston. Of coorse! Maister Garston may be a very smart fellow, an' I believe he is, but wi' a' his smartness he couldna keep a dowg a secret that's capable o' takin' the cup frae auld Ventir.'

For a time this dose of self-encouragement seemed to allay all Davie's fears; still, the mind would soon recall the ghosts of his own creation, and again submit its victim to a process of mild torture. Davie was surprised at himself. So are all men who allow the spirit of speculation to cross its appointed limits and enter the territory of faith. For when our faith is shattered our strength is gone.

The mental billows of this tempestuous sea brought Davie to the trials—for Wednesday had come at last—in a state of under-confidence, which fitted him for the commission of more blunders than fall to the lot of accredited fools. Venture, sharing the timidity of his master, looked nervous and shy, such being the dual penalty incurred by a breach of those mysterious but unfailing laws upon which success depends. They staggered before a storm which had never blown.

The mist, yielding to the rays of the rising sun, gradually unveiled the stern features of the towering hills that hold watch, like unchanging sentinels, over the Fair City of the North. Down the glens, at an early hour, came bronzed shepherds, their wives, sons, and daughters, all *en route* for the trial course, which lay just outside the city. At the outskirts of the town family met family; and some of the looks exchanged there for the first time found fruition at the matrimonial altar in after days.

The indulgence of the prophetic having much to do with the perfecting of our pleasure, there was a perfect Babel of prediction as the rude hill-men journeyed along. Each had his favourite; and the man, or men, who would question the infallibility of one of the " crack hauns," whose reputation had got far beyond the realm of discussion, were regarded as envious in temperament, and lacking in the finer instincts of appreciation and justice. Indeed, by way of giving an admonitory and educational turn to the day's outing, " hazels " were sometimes applied to the cranium of a free-spoken and irreverent critic.

"Davie and Ventir will haud the cup," a plaided septuagenarian, hailing from the Pentlands, opined. "Wait till ye see him at his oot-bye run, and then at his shed. If the auld chap" (meaning Venture) "is in form at a', naething will come near him."

"But what of this great Suffolk dowg, and the yin frae New Zealand? I'm fearin' Davie an' Ventir," argued a less confident brother, "will have the stiffest job they ever had."

"Mon Robin, mon Robin, I'm wae to hear ye. Is't possible to hae seen auld Ventir tak the coorse scores o' times, an' still hae doots? For shame, Robin, for shame, mon! But a week syne"—and a flame lit up the old man's eye—"I saw him on the open hill shed twae lambs off their mother and haud them apairt like a seven-wired fence. Ay, auld Ventir will keep the cup."

What reply, if any, Robin intended to make it is impossible to say, for, just as the last word of a favourite prophecy fell from the old man's lips, a familiar hand was laid upon his shoulder, thus arresting the flow of conversation.

Turning with an agility that was highly inconsistent with his years, and recognising Mr. Blacklock Garston, whom he had known as a boy, the firm-set features of the veteran relaxed, and a film of feeling lent softness to his eye as he exclaimed:

"Mon, Maister Garston—for I'm quite certain I'm no mista'en, ye're sae awfu' like yer mother aboot the een an' yer faither aboot the broo—I'm richt prood to see ye. Peggy an' a wheen mair frae the Glen are comin' on ahint there, nae doot discussin' the price o' stirks an' the next likely mairrage—but I ken their yae great desire is to shake hauns wi' yersel', Maister Garston. Excuse ma feelin's"—the worthy shepherd being slightly overcome—"but, mon, ye lean on yer stick jist like yer faither when he used tae rin auld Whaff, the best dowg I ever saw. Ay, mon, yer faither was the truest and jolliest frien' I ever had."

That Mr. Garston was pleased to meet the old friends

of his boyhood his whole manner clearly showed. The wave of emotion caused by the sympathetic swelling of two hearts brought together after a long separation—this having spent itself, Mr. Garston addressed himself to his friend by asking who, in his opinion, would carry off the cup.

"Auld Ventir, Maister Garston, nane but auld Ventir."

"A great dog, doubtless, but be prepared for surprises."

"What! ye dinna mean to say that that vow ye made aboot lowerin' the pride o' the sooth men is to be brocht into practice the day?"

Again laying his hand on the old shepherd's shoulder, and looking into his eager face, Mr. Garston said:

"Andrew, something, the nearest to old Whaff you ever saw, will take the course to-day. Believe me, the cup goes north to-night."

"Never, Maister Garston, never!"

"We will not argue the matter, Andrew; time and talent will tell. As I must rush to keep an appointment, kindly arrange with Peggy and all the Glen friends to join me at lunch in the field-tent at one o'clock."

When he had left, Andrew looked at his companion and said, "He means it." Then taking off his hat and driving his fingers through his thick grey locks, as if to stimulate the brain to a fresh and greater exertion, but failing to produce anything of a more encouraging character, the Knight of the Crook rested his chin on his breast, and slowly articulated: "He ondootedly means it."

Having meditated for a few moments, he raised his head, looked dreamily at his surroundings, and then, as if about to witness the enactment of a sad fate, moved towards the park in which the trials had to be held.

At the entrance gate all was bustle and excitement. The shepherds exchanged greetings; the dogs held close to the heels of their masters, a youthful first-year competitor now and again taking liberties on the green sward seldom indulged in by his seniors. Spectators crowded in, many asking the while whether the New Zealand and

The Shepherd and His Dog 165

Suffolk dogs had arrived, and in which direction lay the best points of vantage. Ladies fair—blessed (and unblessed) strangers to the tearing toil of the restless world—were there, accompanied by their male protectors and *cicerones*, whose only acquaintance with the shepherd and his calling was through the medium of literature. Philosophy, science, art, and the empire's industries were represented on the field that day. For once age and devotion had that priority which their nature suggest—the successors of the watchers on the plains of Bethlehem stood first.

The little knots of critics and admirers located with all the exactness of a topographical map the outstanding favourites. David Garlow and Venture commanded by far the largest crowd. Then came the New Zealand and Suffolk champions—their audiences being more select than numerous.

At a distance stood a small, square, kindly-faced man, with a dog that looked a reflex of himself. Nobody seemed to know him; and his manner, likewise the manner of his dog, indicated a retiring shyness. Possibly he had just dropped in to witness the contest.

The judges, three skilled men and true, took their places: the Director of the Course called up No. 1, and so the quill was upon the page of history.

No. 8 having been announced, David Garlow and Venture stepped up to the starting-post amid thunders of applause. The great veteran led away beautifully, but there was hesitation in his command, which feature marked his whole performance. He was always holding up for the emergency that never came. Still at the close of his run it was felt that the cup was his. True, he had done better; David knew it and censured himself; but no dog could hope to equal his second best. Enthusiastically, the crowd sought to crown Davie before his regal rights had been established.

The New Zealand and Suffolk dogs ran and made brilliant appearances, but still the cup lay with Davie.

"Ah, Mr. Garston," said his friend Andrew, " ye're no

to manage efter a'. The glory o' the sooth is safe in the hauns o' Davie an' Ventir."

" Have patience, Andrew, have patience. You know, miracles never come early in the day," laughingly replied the genial fancier.

" No. 17," shouted the Course Director. Slowly there emerged from the crowd the small, square, kindly-faced man, accompanied by his little, mild-featured collie.

" Who's this ? " several asked.

" Magnus Drever, frae Orkney, wi' Ronald," a stripling, who had been consulting the catalogue, replied.

" Weel, Magnus, my man, ye've come a lang road for little oo," a voice remarked.

" Say for the cup, Andrew ; say for the cup, and you will be no false prophet," quietly retorted Mr. Blacklock Garston.

" Maister Garston, I wunner to hear――" Andrew's lips suddenly became sealed ; then they parted, giving spacious emphasis to the world of wonders that now opened before him.

Ronald had compassed the course and taken command of his sheep in faultless style. Moving up to his master with his charge, Ronald revealed an art so perfect and bewitching that spectators felt too deeply to applaud. Praise loudly uttered is sometimes profanity.

With a peculiarly subdued air of confidence, Magnus stepped out to meet and co-operate with his dog for the purpose of effecting a shed. By two lightning moves Ronald drew up the two marked sheep and held them like a wall of iron.

" But what of the pen ? " a prophet of misfortune whispered.

Already Magnus had opened the gate, and, unlike his rivals, stood there, allowing Ronald to take the full burden of penning. Not a few of the experts felt that his confidence had actually become dangerous.

Ronald knew his powers, and bravely applied them. In a minute and a half he had his lot inside, his master never having moved. Even now Magnus stood holding the gate wide open. Like a giant Ronald held the mouth of the pen.

"Let go the single sheep," cried one of the judges. Nobly Ronald buckled to this, the final test. Not an inch of ground would he surrender; in twenty seconds he was master; the sheep had to obey. Responding to a whisper from his master, Ronald crept up to his charge, making it back towards Magnus until the gentle little Orcadian, who still occupied his original position at the pen, laid his hand upon it.

Pent-up feelings could no longer be restrained. Cheer after cheer rent the air, the more impatient youths rushing into the enclosure to congratulate the victor. Even the judges forgot the impartial propriety of their office and joined in the applause.

Offering his hand to Mr. Garston, Andrew said in a husky voice that told of both pain and pleasure :

"It's away, Mr. Garston, it's away. Ye were richt— aye, ye were richt. Ronald is jist yer faither's auld Whaff come to life again."

That night, in a humble turf-roofed cot in Stronsay (Orkney) a little tidy white-haired old woman received and read a telegram, then knelt down and prayed. She was Magnus Drever's widowed mother.

SHEPHERDS' DOGS IN CHURCH

By The Author

Shepherds' dogs in kirk is no rare sight in Scotland. We do not hear of whip and tongs being in requisition for the maintenance of order as in some parts of the country, but to judge from what we find in *Notes and Queries* (1876),[1] as given below, the behaviour of our northern canine friends on these occasions is not always perfect.

"An Edinburgh minister was performing service one Sunday in a remote country kirk, where dogs formed no inconsiderable part of the congregation. It is the custom

---

[1] By permission of Editor.

of the Scotch Kirk for the assembled worshippers to stand while the blessing is pronounced. When the minister, however, rose for that purpose at the end of the service, he perceived, to his surprise, that his hearers all remained seated. He looked around for some little time with an expectant eye, but no one moved. At last the clerk, with the view of relieving the honest gentleman's embarrassment, turned up his head from his desk below, and bawled out, 'Say awa', sir, it's joost to cheat the dowgs!' It had been found that the dogs, imagining the service to be concluded when the congregation stood up at this crisis, always prepared for their own departure, and disturbed the solemnity of the occasion by various canine noises and shufflings; they had, therefore, to be circumvented by the people keeping their seats while the benediction was given."

Another correspondent adds: "I have a vivid recollection of an anecdote which my father used to relate, nearly if not quite half a century ago, with regard to dogs being taken to public worship in Scotland. In a rural kirk where this was the practice, the shepherds' dogs were permitted to occupy the gallery over their masters' heads, where they remained during service time, and, it is fair to suppose, conducted themselves in an inoffensive manner; but on one occasion, presumably that of a larger assembly than usual, a strange dog was introduced among them. This was a signal for a general commotion upstairs, which terminated by the sudden bolting of the intruder over the front of the gallery into the body of the church, and as speedily out of it by the door, pursued by the same route in his headlong exit by the whole dog congregation. This amusing anecdote acquires peculiar interest from having been originally related by the celebrated Edward Irving; and the occurrence, if I am not mistaken, took place under his own eyes, and probably in consequence of his popularity."

Another anecdote is related by C. F. S. Warren: "In 1839 a relation of mine was fishing on the Whitadder when a small building attracted his attention, and he

The Shepherd and His Dog

asked a shepherd, 'Pray, is that a kirk? it looks very small'; to which the shepherd answered, 'Aye, aye; but it's no sae sma'; there's aboon thirty collies there ilka Sabbath.'"

The Rev. J. E. Vaux, in *Church Folklore* (1902),[1] gives the following account of shepherds and their dogs in a chapel in Ireland :—

"About twenty years ago I was in Connemara salmon-fishing. The first Sunday the landlord of the hotel where I was staying kindly offered me a seat in his car to convey me to a chapel on the bog, three or four miles off, for the midday Mass. I gladly accepted the lift. The chapel was of the most primitive kind. The floor was but of beaten clay. When I entered, the altar-rail was closely packed with worshippers, who, I presume, were all shepherds. There was only one pew, which belonged to the 'quality,' *i.e.* the landlord and his family. I preferred to kneel alongside my attendant 'ghillie' (to use a Scotch term) who was there. There were a dozen dogs at least in the chapel, several of them sitting behind their masters, who were kneeling at the altar-rails. One of these sheep-dogs attracted my attention. He sat most quietly through the earlier portion of the service. As soon as the creed had been recited, and the celebrant turned round to deliver the sermon, the dog looked up as much as to say, 'Oh! sermon time, all right,' and having, dog fashion, walked round three times, curled himself up for a comfortable sleep. The sermon, which did not last more than ten minutes, being over, the dog woke up, and sat on his tail behind his shepherd master until the service was over. There was something so deliciously human about this, that I have never forgotten it. I have described the incident exactly as it happened, without the slightest exaggeration."

[1] By permission of Messrs. Skeffington & Son.

SHEPHERDS' DOGS EXPELLED FROM CHURCH

By the Venerable ARCHDEACON THOMAS, F.S.A.

Some of the stories told by the late Dean Ramsay, in his *Reminiscences of Scottish Life and Character*, of the sagacity of collie dogs, must, to judge from certain mementos, have had their amusing as well as ridiculous counterparts in the Principality, only they have lacked the pen of the witty dean to chronicle them. Following their masters through the labours of the week, they did not see why they should not share their Sabbath observances; but they had their own notions of the proper length of such indulgences, and they had their own ways of making their opinions known. Neither were they altogether free from the clannish pride and partisanship of their owners; indeed, it was no uncommon thing for them to start up in vigorous assertion of their offended dignity, and that at moments and in places highly inopportune; and many a stout heart that would have collared his offending fellow-man, kept at a prudent distance from the uninviting teeth of the too faithful companion. Still certain unpleasant duties had to be performed, and a timely invention came to the aid of the disconcerted churchwarden. The illustration given (Fig. 4) shows very well the form of the instrument both at rest and in motion, and its character has become familiar to us in another use, under the name of "lazy tongs." Some of the joints, including the handle, have been lost from the present specimen; but the handle was not unlike the forceps or catching end, which was in some cases (as at Gyffylliog) lined with nail-heads or small knobs to make the grip more secure as well as more cautionary. When no convenient pew could shelter the offender, and no amount of snarling could any longer ward off the certain, not to say ignominious, expulsion of the culprit, the dog-tongs had only to be quietly taken off the seat on which they lay so innocently, and the handles brought quickly together,

Copyright, Rev. Morris Griffith.   FIG. 1

IRON DOG-TONGS USED TO DRAG PUGNACIOUS SHEPHERDS' DOGS FROM PENMYNYDD CHURCH, ANGLESEY

The wooden handles are new; the fangs either almost worn out or have been purposely mutilated; length when extended, 4 feet 6 inches.

Copyright, Rev. J. J. Ellis.   FIG. 2

WOODEN DOG-TONGS IN LLANEILIAN CHURCH, ANGLESEY

FIG. 3
DOG-TONGS IN BANGOR CATHEDRAL

when out shot the jointed folds and arms, and in an instant seized the helpless wretch around the neck or leg, and without danger or ceremony extruded him from the place. The usefulness of such an instrument must have been very great when dogs were more in the habit of attending church than they happily now are, and when it was even necessary to appoint an officer to see to their proper conduct, or, if necessary, their summary exclusion. There was one occasion on which the presence of a dog

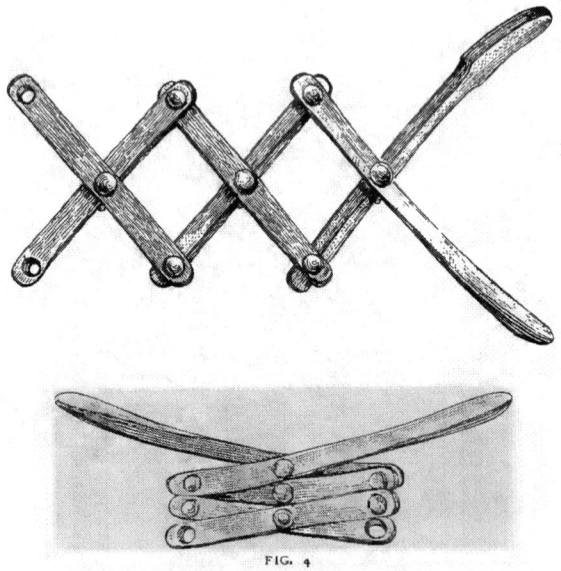

FIG. 4
DOG-TONGS, CLOSED AND OPEN

was held to be specially ominous, for Pennant tells us that " Among the Highlanders, during the marriage ceremony, great care was taken that dogs should not pass between the couple to be married " (Brand's *Popular Antiquities*, ii. p. 170). Whether such a custom prevailed also in the Principality does not appear, neither are we told the reason of the precaution ; but may it not have been interpreted as an omen that there would be more love for the old dog than for the new wife?

The tongs here illustrated are from Clodock Church,

The Shepherd and His Dog

in Herefordshire, and were exhibited by the Rev. C. L. Eagles in the Temporary Museum at Abergavenny, in 1876. A similar pair, but more perfect, from Llanynys Church, Denbighshire, was exhibited by the Rev. John Davies, vicar, at the Wrexham Meeting in 1874. Another, as already mentioned, existed in Gyffylliog Church in the same county.[1]

Copyright, A. Coates.

PLATE I

DOG-WHIP, PATEN, AND WOODEN COLLECTING-BOX
IN BASLOW CHURCH, DERBYSHIRE

OTHER METHODS OF EXPULSION

By The Author

Another way of expelling such canine intruders was by whipping. In old days it was the custom in various parishes in England, in the Isle of Man, and in Wales to appoint a dog-whipper, to keep the shepherds' dogs out of churches. It is interesting to note that in Thomas Wright's *Dictionary of Obsolete and Provincial English* (1857) he has

---

[1] Archdeacon Thomas, in a letter to the author, adds: "I myself have seen dogs accompany the shepherds to church in 1851, in the old parish church of Llandrindod, but they behaved very well, and there was no need to use the tongs with them. Besides those tongs mentioned in the paper at Gyffylliog and Llanynys, there are also others at Nantglyn (these are three adjoining parishes in the Vale of Clwyd), and a pair preserved in Bangor Cathedral, but I forget whence they came; and another pair at Clynnog Fawr, made of iron, and dated 1815. The arms of some of these tongs extend out six or seven feet."

"Dognoper = a beadle" (Yorks), also "Dog-whipper = a beadle" (North). Mr. Arthur Finn tells me of an entry in the churchwardens' book of Lydd, Kent, dated 1520: "I first paid to Robert Foule for keeping the dogs oute of the church and for the last year gotte throng, xx d." And Archdeacon Thomas sends me the following from the vestry books of the parish of Llandrindod :—

		£	s.	d.
1811.	Wm. Thomas, the Dog Whipper .	0	8	0
1818.	For a lash to whip the dogs . . .	0	0	4
1819.	Eleanor Thomas, Dog Whipper, ½ yr.	0	4	0
	(She succeeded her husband.)			

Whipping the dogs out of church is said to have been the custom at Baslow, Derbyshire, until the beginning of the last century. And a dog-whipper was regularly appointed and used this whip: "The throng of the whip is about three feet long, and is fastened to a short ash stick, round the handle of which is a band of twisted leather." (*See* Plate I.) Plate II. is from a life-size portrait of "Old Scarleit" which hangs in the nave of Peterborough Cathedral. His dog-whip is seen thrust through his belt. "His office by theis tokens you may know."

At Broseley, in Shropshire, a similar official is said to have been familiarly called the "Whipper-in and the Scouter-out," and Halliwell (1860) has: "Dognoper = the parish beadle" (Yorkshire).

Notes and Queries (1875) supplies information on whipping dogs out of church from the following church registers :—

Trysull, Staffordshire (17th April 1725); Claverley, Shropshire (25th August 1659); Chislet, Kent; Peter's Church, Herefordshire; Louth (from 1550 at intervals till 1705); St. Mary's, Reading (1571); Smarden, Kent (1576); Battle Church (1633): Yolgrave Church (1609, 1617); Ogbourne St. George, near Marlborough (1632, 1633, 1639); East Witton, Yorkshire; Goosnargh, Lancashire (10th April, 1704); St. Mary-le-Bow, Durham (6th April 1722); Kirton-in-Lindsey, 1817.

The Shepherd and His Dog 175

The late Rev. J. Eastwood, in his *History of Ecclesfield*, Co. York (1862), speaks of the dog-whipper as being still known under the name of "the dog-noper."

Mr. R. Butterworth, one of the churchwardens of Ecclesfield

YOV SEE OLD SCARLEITS PICTVRE STAND ON HIE
BVT AT YOVR FEETE THERE DOTH HIS BODY LYE
HIS GRAVESTONE DOTH HIS AGE AND DEATH TIME SHOW
HIS OFFICE BY THEIS TOKENS YOV MAY KNOW
SECOND TO NONE FOR STRENGTH AND STVRDYE LIMM
A SCAREBABE MIGHTY VOICE WITH VISAGE GRIM
HEE HAD INTERD TWO QVEENES WITHIN THIS PLACE
AND THIS TOWNES HOVSE HOLDERS IN HIS I IVES SPACE
TWICE OVER: BVT AT LENGTH HIS OWN TVRN CAME
WHAT HE FOR OTHERS DID FOR HIM THE SAME
WAS DONE: NO DOVBT HIS SOVLE DOTH LIVE FOR AYE
IN HEAVEN: THO HERE HIS BODY CLAD IN CLAY

PLATE II

(Yorks, 1910), who was contemporary with the Rev. J. Eastwood, remembers these dog-nopers being talked of. "Nope" in northern dialect means a knock on the head. In 1785, in W. Hutton's *Bran New Wark*, p. 157, we find: "In some churches the sidesmen gang about with staves and give every sleeper a good nope."

SHEPHERDS' DOGS AND SHEEP-STEALING

By JAMES HOGG ("The Ettrick Shepherd"), 1772-1835

The stories related of the dogs of sheep-stealers are fairly beyond all credibility. I cannot mention names, for the sake of families that still remain in the country; but there have been sundry men executed,[1] who belonged to this district of the kingdom, for that heinous crime, in my own days; and others have absconded, just in time to save their necks. There was not one of these to whom I allude who did not acknowledge his dog to be the greatest aggressor. One young man in particular, who was, I believe, overtaken by justice for his first offence, stated, that after he had folded the sheep by moonlight, and selected his number from the flock of a former master, he took them out, and set away with them towards Edinburgh. But before he had got them quite off the farm, his conscience smote him, as he said (but more likely a dread of that which soon followed), and he quitted the sheep, letting them go again to the hill. He called his dog off them, and mounting his pony he rode away. At that time, he said, his dog was capering and playing around him, as if glad of having got free of a troublesome business; and he regarded him no more, till, after having rode about three miles, he thought again and again that he heard something coming up behind him. Halting, at length, to ascertain what it was, in a few minutes there comes his dog with the stolen animals, driving them at a furious rate to keep up with his master. The sheep were all smoking, and hanging out their tongues, and their guide was as fully as warm as they. The young man was now exceedingly troubled, for the sheep having been brought so far from home, he dreaded there would be a pursuit, and he could not get them home again before day. Resolving, at all events, to keep his hands clear of them, he corrected his dog in great wrath, left the sheep once more, and, taking the collie with him, rode off a second

---

[1] Under an Act long since repealed.—[*Author's Note.*]

time. He had not ridden above a mile, till he perceived that his assistant had again given him the slip ; and suspecting for what purpose, he was terribly alarmed as well as chagrined ; for daylight now approached, and he durst not make a noise calling his dog, for fear of alarming the neighbourhood, in a place where they were well known. He resolved, therefore, to abandon the animal to himself, and take a road across the country, which he was sure the other did not know, and could not follow. He took the road, but, being on horseback, he could not get across the enclosed fields. He at length came to a gate, which he shut behind him, and went about half a mile farther, by a zigzag course, to a farm-house, where both his sister and sweetheart lived ; and at that place he remained until after breakfast time. The people of this house were all examined on the trial, and no one had either seen the sheep or heard them mentioned, save one man, who came up to the aggressor as he was standing at the stable door, and told him that his dog had the sheep safe enough down at the Crooked Yett, and he need not hurry himself. He answered that the sheep were not his—they were young Mr. Thomson's, who had left them to his charge, and he was in search of a man to drive them, which made him come off his road. After this discovery, it was impossible for the poor fellow to get quit of them ; so he went down and took possession of the stolen drove once more, carried them on, and disposed of them ; and, finally, the transaction cost him his life. The dog, for the last four or five miles that he brought the sheep, could have had no other guide to the road his master had gone but the smell of his pony's feet.

It is also well known that there was a notorious sheep-stealer in the county of Midlothian, who, had it not been for the skins and heads, would never have been condemned, as he could, with the greatest ease, have proved an *alibi* every time suspicions were entertained against him. He always went by one road, calling on his acquaintances, and taking care to appear to everybody by whom he was known, while his dog went another way with the stolen sheep; and then, on the two felons meeting again, they had nothing more to do than turn the sheep into an associate's enclosure, in whose house the dog was well fed and entertained, and would have soon taken all the fat sheep on the Lothian edges to that house. . . . On the disappearance of her master she lay about the hills and places where he had frequented, but she never attempted to steal a drove by herself, nor the smallest thing for her own hand. She was kept some time by a relation of her master's, but never acting heartily in his service, soon came to an untimely end.

SHEEP-MARKS AND TALLIES

By Habberton Lulham.
THE BADGE OF OWNERSHIP

SHEEP-MARKS AND TALLIES

> Every shepherd tells his tale[1]
> Under the hawthorne in the dale.
> MILTON, *L' Allegro.*

ON SHEEP-MARKING

By THE AUTHOR

THE following from *Curious Church Customs*, by William Andrews (1896), is of special interest, as it gives a

[1] Counts his sheep.

record of ear-marking in England in the seventeenth century :—

"In April 1864, when clearing out a large, roughly made, lidless chest, beneath the tower of the parish church of Luccombe, Somerset, wherein the sexton kept his tools, a considerable store of decaying papers came to light beneath a mildewed parish pall. It is well known to ecclesiologists that up to comparatively recent times a variety of notices, that would now considerably startle demure congregations, were given out in parish churches. . . . But as far as our acquaintance with old parochial documents extends, the Luccombe chest is the only one that has yielded absolute evidence of announcements being made in church in the seventeenth century about strayed cattle. By no means the least interesting of the curious medley of fragments so strangely preserved in this Somerset village was one of which the following is a copy :—

The Clerke shal give notice on Trinitie Sondaye after divine service is ended publickly in the Chuche that one score and three straye sheepe hav bin vounde in David Pugsley his bartone with a clippette in the ye lefte eare. Also that a redde cowe hath bene pinned by the pyndere of East Luckham.

The writing is good, too good probably for the village constable, and is most likely that of the rector or curate. The sheep had doubtless strayed off the closely adjacent Exmoor. The 'Zomerzet' *v* for *f* may be noticed in 'vounde' for 'found.' This paper is not dated, but there can be no doubt that it is of the reign of Charles I."

From a curious *Shepherd's Guide* (the date of which is not given), under the heading of Matterdale (chap. xii.), Southey, in his *Commonplace Book* (1849-1851), quotes some interesting particulars as to the former methods employed in the marking of sheep in the English Lake Country. As this book is not in the British Museum, it may be worth while to give a full account of it. The subject of the book is given as follows :—"The Shepherd's Guide, or a delineation of the wool and ear marks on the different stocks of sheep in Patterdale, Grasmere, Hawkshead, Langdale, Loughrigg, Wythburn,

Legberthwaite, St. John's, Wanthwaite and Burns, Borrowdale, Newlands, Threlkeld, Matterdale, Watermillock, Eskdale, and Wastdalehead."[1]

The original preface says :—" The success this work has met with is sufficient to show the extensive benefit which is likely to result from it. It has not been presented to any sheep-breeder who has not considered it of the greatest importance. My object is to lay down a plan by which every man may have it in his power to know the owner of a strayed sheep, and to restore it to him, and at the same time, that it may act as an antidote against the fraudulent practice too often followed—in a word, to restore to every man his own.

"I considered that the best mode of representing the wool and ear marks would be to have printed delineations of the animals, on which the respective marks might be laid down, and to which the printed description would serve as an index. Accordingly the book consists of fourteen chapters of prints, filling eighty-four pages, with three couple of sheep in each, each couple being numbered."

As above intimated, Southey selects the chapter on Matterdale, from which he quotes the following detailed descriptions of the marks employed, with which he couples the name of the farmer to whom the sheep in question belonged. Thus we have :—

"No. 12. William Calvert, Esq., Wallthwaite.
   Bitted far ear, old sheep. M on the nearside; hogs, full cripping across each buttock, and no letter.
No. 17. John Sutton.
   Cropped, and muck-forked on the far ear; under fold bitted on the near; a red stroke over the fillets of the near side, the form of a grindstone handle.
No. 23. John Brownrigg, Matterdale End.
   Cropped far ear, bitted near; a red stroke on the top of the shoulder; J. B. on the near side.

The ear-marks are what are most depended on, because

---

[1] The names of the book's two authors are given as William Mounsey and William Kirkpatrick, and the book is described as being on the plan "originally devised by Joseph Walker." It was octavo in size, and printed (without date) by William W. Stephen at Penrith.

they cannot so easily be got rid of. The ear is either cropt, under or upper halved, under key-bitted or upper-holed, muck-forked, or clicking-forked, marked with a three-square hole, etc. ; and these marks are varied, by being either on the cropt or otherwise entire ear. The other marks have all their technical names."

Southey adds : " The copy before me is one which my brother T. has borrowed from a neighbour. It is neatly bound in red sheep ; and has pasted in it a printed paper with these words, ' Newland's Public Book.' The sheep are coloured according to the description, and a blank in the engraving left for the ears of one of the couple."

In old days ear-marking was the custom in the Isle of Man and was called " cowrey keyrragh " or " bein er y chleaysh." Now it is usual for the sheep to have initials painted on their fleeces, or when the wool is dark, as with the laughtans, to brand the horns. Their noses were never branded with a hot iron. The following in respect to unmarked sheep in the island is interesting :—

SAINT COLUMBA'S EVE, 9*th June*.[1]

According to the " Ordinances " of the Isle of Man, *anno* 1510.

" The fforester or his deputy ought to go forth on St. Collum Eve through the fforest, and ride to the highest hill-top in the Isle of Man, and there blow his horn thrice ; this done, to range and view the fforest, and on the third day to go forth and take such company with him as he shall like, to see what sheep he findeth unshorn. If he finde any, he ought to take them with his dogge, if the said sheep be not milk sheep, to shear them and to take the fleece to himself, and to put a private mark upon said sheep, to use all he finds within the precincts of the fforest so at the time, to the intent that if any of the said sheep be found the next year by the same fforester, he to certify the comptroller and receiver of the same, that they may be recorded in the Court Rolls and so priced and sold to the Lord's best profit, etc."

An old Manx man says that the forester's mark in his young days was to cut a tiny strip of skin almost off the tail underneath, and to twist it tightly into something like

---

[1] The day was altered to 21st of June by statute of 1748, chap. vi.

a tag. The office of forester was abolished by the Isle of Man Disafforesting Act of 1860, sec. 16.

With regard to Wales, the Rev. Morris Griffith, writing from Anglesey, says:—"The branding of lambs with a hot iron on the nose ceased here and in Carnarvonshire twenty years ago. The custom which prevails in the Snowdonian range, where thousands of sheep graze, is to make a tar-mark on the lambs when they are taken from the mountains in the spring, and to mark their ears when they are collected for shearing purposes in June. Every farm has its own tar- and ear-mark, so that they may be able to identify their sheep. Thus, for ear-mark—

Left ear.   Right ear.

The tips of both ears are cut off, and the knife is drawn lightly under the left ear for the other mark. The above is the specimen of the marks of a farm in Anglesey. The burning of initials on their horns is a common custom, but all sheep have not horns; Welsh sheep, as a rule, have very short ones. After shearing, an iron brand is steeped in hot pitch and initials stamped on their bodies, different farms stamping different parts."

In Ireland ear-marking still obtains, although the tendency (in the case of blackfaces) is to brand the horns. Then, on being clipped, the initials of owner or place are described in tar on some part of each sheep. The growth of more tender feelings towards the animal world is operating against ear-marking.

LAMB-BRANDING IN SKYE

By ALEXANDER SMITH, 1865

Morning broke forth gloriously — not a speck of vapour on the Cuchullins; the long stretch of Strathaird wonderfully distinct; the loch bright in sunlight. . . .

We went up the glen, and as we drew near the "fank" we saw a number of men standing about, their plaids thrown on the turfen walls, with sheep-dogs couched thereupon; a thick column of peat-smoke rising up smelt easily at the distance of half a mile; no sheep were visible, but the air was filled with bleatings—undulating with the clear plaintive trebles of innumerable ewes and the hoarser *baa* of tups. When we arrived, we found the narrow chambers and compartments at one end of the "fank" crowded with lambs, so closely wedged together that they could hardly move, and between these chambers and compartments temporary barriers erected, so that no animal could pass from one to the other. The shepherds must have had severe work of it that morning. It was as yet only eleven o'clock, and since early dawn they and their dogs had coursed over an area of ten miles, sweeping every hill-face, visiting every glen, and driving down rills of sheep towards the central spot. Having got the animals down, the business of assortment began. The most perfect ewes—destined to be the mothers of the next brood of lambs on the farm—were placed in one chamber; the second best, whose fate it was to be sold at Inverness, were placed in a congeries of compartments, the one opening into the other; the inferior qualities—*shots* as they are technically called—occupied a place by themselves: these were also to be sold at Inverness, but at lower prices than the others. The fank is a large square enclosure. The compartments into which the bleating flocks were huddled occupied about one-half of the walled-in space, the remainder being perfectly vacant.

One of the compartments opened into this space, but a temporary barrier prevented all egress. Just at the mouth of this barrier we could see the white ashes and the dull orange glow of the peat-fire, in which some half-dozen branding-irons were heating. When everything was prepared, two or three men entered into this open space. One took his seat on a large smooth stone by the side of the peat-fire, a second vaulted into the struggling mass of heads and fleeces, a third opened the barrier slightly,

Sheep-Marks and Tallies

lugged out a struggling lamb by the horns, and consigned it to the care of the man seated on the smooth stone. This worthy got the animal dexterously between his legs so that it was unable to struggle, laid its head down on his thigh, seized from the orange glow of the smouldering peat-fire one of the red-hot heating-irons, and with a hiss and a slight curl of smoke drew it in a diagonal direction across its nose. Before the animal was sufficiently branded the iron had to be applied twice or thrice. It was then released, and trotted bleating into the open space, perhaps making a curious bound on the way as if in bravado, or shaking its head hurriedly as if snuff had been thrown into its eyes. All day the branding goes on.

The peat-fire is replenished when needed; another man takes his seat on the smooth stone; by two o'clock a string of women bring the dinner from the house, and all the while young M'Ian sits on the turfen wall, notebook in hand, setting down the number of the lambs and their respective qualities. Every farmer has his own peculiar brand, and by it he can identify a member of his stock if it should go astray. The brand is to the farmer what a trade mark is to a manufacturer. These brands are familiar to the drovers, even as the brands of wine and cigars are familiar to the connoisseurs in those articles.

The operation looks a cruel one, but it is not perfectly clear that the sheep suffer much under it. While under the iron they are quite quiet—they neither bleat nor struggle—and when they get off they make no sign of discomfort save the high bound or the restless shake of the head already mentioned (if indeed these are signs of discomfort, a conclusion which no sheep farmer will in any way allow). In a minute or so they are cropping herbage in the open space of the "fank," or if the day is warm, lying in the cool shadows of the walls as composedly as if nothing had happened. Leaning against the "fank" walls, we looked on for about an hour, by which time a couple of hundred lambs had been branded.[1]

[1] Lamb-branding as described by Alexander Smith seems to be a thing of the past. —[*Author's Note.*]

SHEEP-MARKS

By the Rev. Thomas Mathewson, 1909

These marks were made upon the animal's ears. Each family knew their sheep by the mark, which descended to the youngest son. I have in my possession the only existing copy, as far as I am aware, of the sheep-marks, with the names of the persons to whom they belonged, of the parish of Northmavine, Shetland, more than a hundred years ago. Dr. Hibbert, writing in 1822, says : " There is a code of sheep laws, preserved in Debes's *Description of Feroe*, which is dated Opslo, A.D. 1040, being addressed from Hagen, Duke of Norway, and son of King Magnus, to the Bishop of Feroe and Mr. Sefvort, Provincial Judge of Shetland, named here Hetland. From the tenor of this sheep ordinance it evidently relates to an enclosed state of the country. The laws corrected the grievances that arose from unmarked, stray, and wild sheep, from a clandestine marking of lambs, from trespasses upon fields or enclosures, from keeping a superfluous number of sheep-dogs, and from sheep being injured or destroyed by dogs not properly trained to their office."[1] In a previous chapter Dr. Hibbert writes : " As the seizure of sheep took place by means of dogs, it was necessary for the preservation of individual property that no capture should be private. Every proprietor in claiming his share of a promiscuous flock had a particular mark of his own that was formed by various kinds of incisions, which were inflicted on one or both of the animal's ears ; these received such names as a shear, a slit, a hole, a bit out of the right or left ear, before, behind, or from the top. In this way an infinite variety of private marks was devised, but none of these could be lawfully used without the sanction of the bailiff of a district or civil officer, whose duty it was to insert in a public register a descriptive account of all the tokens which any individual wished to adopt for the recognition of the particular share which he had in a joint

---

[1] *Description of the Shetland Islands*, pp. 471-2.

flock of sheep. It was, therefore, a proper regulation that the marking of sheep should be a public act, and that no property could be thus claimed but in the sight of a whole district. The period appointed for marking lambs was when all the proprietors of a flock were assembled for the purpose of 'rueing,' or tearing off with the hand the wool from sheep after it had naturally begun to loosen; this was about the middle of May, or near midsummer.

"Thus there was a law that no one should mark lambs or rue sheep where there are different owners in the flock but in the sight of sufficient witnesses, under the pain of ten pounds scots for the first offence, and of double the amount for the second, and for the third fault of being reputed and punished as thieves. The time of marking and rueing is still publicly proclaimed, and on the day fixed all the men of a district turn out and drive their common flock, without any preparation of washing, into rude enclosures, named *punds* or *crues*. If the punding be delayed too long the sheep become so wild that they are hunted down and taken by dogs; but when at last they are secured within the *crues* the civil officers (who were in former days the bailiff and ranselmen of a district) appear as arbiters of all disputes."[1]

Mr. Shirreff, writing in 1814, has given a curious specimen of the register of a sheep-mark as taken from the parish records of Orkney, where a custom nearly similar to the Shetland practice prevailed: "I, John Gillies, baron-bailie of the parish of Orphir, hereby grant warrant to Edward Wishart, in Mill of Claistran, to assume and use the sheep-mark following, as the same is recorded in the register of sheep-marks on the 4th day of July 1770 years in the name of John Flett in Skelbister, viz. The crop of the right lug and a bit behind, a rip in the left lug and a bit before, and the tail off."[2]

---

[1] *Description of the Shetland Islands* (1822), pp. 438-9.
[2] *Agricultural Survey of Orkney*, by John Shirreff, p. 132.

EAR-MARKING IN SHETLAND
By Dr. Jakobsen, 1909

Old Shetland Name given to Mark.	Description.
Middled or grind	Square bit cut out of top of lug. Lit. = " gate."
Fidder	An incision.
Hingin' widder (hingin' fidder)	Opposite of fidder.
Crook	An angular piece cut out.
Bit	What we should call a " bite."
Shule	Triangular bit, generally at top.
Strae-draw	Narrow cut down alongside of lug.
Stoo	Top off.
Rit i de stoo	Top off and slit down.
Getskerdand a hole	A hole in connexion with a hole.
Gongbit	Two " bits " on opposite sides.[1]

[1] I am much indebted to Dr. Jakobsen for these names. A full list will appear in his learned and valuable *Shetland Norn Dictionary*. Parts I. and II. are already published.—[*Author's Note.*]

TALLY-STICK REGISTERS

By Edward Lovett, 1909

This method of recording numbers by notches on tally[1]-sticks is still practised by a few shepherds of the old school, and I have obtained specimens of flock tallies as well as lamb tallies. These are cut either on squared lengths of wood about half an inch wide and eight or nine

Fig. 1.
A LAMB TALLY FROM WORCESTERSHIRE
23 red notches (doubles), 46 lambs.

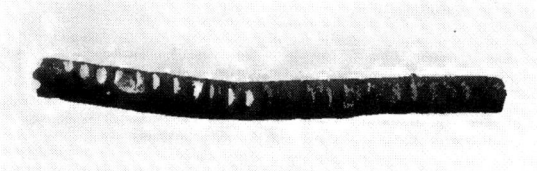

Fig. 2. (Reverse Side.)
11 plain (singles), 12 black (triplets), 47 lambs.
The total, 93, is marked on the ends.

inches long, or on natural round sticks of about the same size, and with the bark left on. In the lamb tallies used for recording the number of lambs born in the season an ordinary notch denotes doubles or twins, a short notch or dot a single, and perhaps an extra long notch trebles. (In

---

[1] (F.—L.) F. *taille*, a notch, cut; L. *talea*, a slip of wood. The final *y* in tall-y is due to the frequent use of F. *taillé*, pt.p., to signify "notched."—[*Author's Note.*]

Worcestershire the doubles are similarly marked by ordinary notches, but the singles and trebles by black and red coloured notches. See Figs. 1 and 2.) By this simple method a shepherd can quickly record and "tot up" the number of his lambs.[1]

The flock tallies are used when the lambs are old enough to leave the ewes, and the time has come for dividing up the flock. In dividing, the animals are separated by twenties, or by the "score" (*i.e.* by the scratch or notch). After five "scores" have been made the fifth notch is continued either over the edge of a squared tally or further round a natural bark-stick-tally than the other notches, so that the hundreds on the tally can be read off simply and easily. Any odd animals are marked by smaller notches at the end of the row, so that, for a new flock of say 613 sheep, the completed tally would show six sets of five notches each, followed by thirteen smaller notches, thus :—

FIG. 3.

Fig. 4 shows an actual flock tally recording 506 sheep. One of the old shepherds made a very remarkable tally for me, saying that his grandfather used one like it. It consists of a piece of natural wood with the bark on, about one inch in diameter and six inches long. This is hollowed out, and the ends stopped with two bits of cork. In this wooden bottle are placed small pebbles, each one representing a score of sheep, and the old sheep are notched upon the bark in the same way as on the ordinary tally. In this tally a flock of 613 sheep would be recorded by thirty small pebbles and thirteen notches. In the same locality—Burpham, near Arundel—tallies are still used for other purposes than those described above. For example, when a man buys lambs for feeding up, he will have a

---

[1] Curiously enough, they can guess about how many to expect, so the stick is evenly filled, for I am told that threes and ones invariably equal the twos in a flock.—E. LOVETT.

record of the original number in stock, and will have a tally record of the deaths.

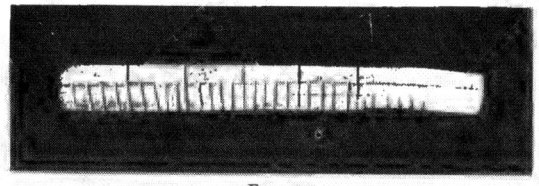

FIG. 4.

NOTCHES AND NICKS
By THE AUTHOR

Old Sussex shepherds tell me that when they were boys they used to play cricket on the Downs, and register the scores by cutting notches on sticks. This method was once general, and accounts for the survival of the word "notch" in the sense of a "run." In accounts of modern cricket matches we may constantly read that So-and-so "scored so many notches to his credit." Dr. E. Cobden Brewer, in his *Dictionary of Phrase and Fable*,[1] remarks that "tallies" used to be called "nick-sticks." Hence to make a record of anything is to nick it down.

"In the nick of time," just at the right moment. The allusion is to tallies marked with nicks or notches. Shakespeare has, "'Tis now the prick of noon,"[2] in allusion to the custom of pricking tallies with a pin, as they do at Cambridge University still. If a man enters chapel just before the doors close, he would be just in time to get nicked, or pricked, and would be "in the nick of time." Halliwell gives the following :—

"*Nick-stick.*—A tally, or stick notched for reckoning (northern).

"*Nicky.*—A faggot of wood (west).

"*Nick.*—Used in the proverbial expression "to knock a nick in the post," *i.e.* to make a record of any remarkable event. This is evidently an ancient method of recording. Similarly we have 'in the nick of time,' *i.e.* just as the notch was being cut ; in the nick exactly (northern)."

---

[1] Cassell & Co.   [2] *Romeo and Juliet*, ii. 4.

THE SHEEP-COUNTING SCORE

By Walter Skeat, 1910

An old farmer in the county of Sussex, rather more than half a century ago, observed that a drover, from whom he wished to purchase some sheep, made use of an unusual set of numbers (which began with *een*, *doit*,[1] *tree*, and ran up to twenty) by which to reckon them. Feeling a very natural curiosity in this discovery, he determined, if possible, to ascertain their origin. The numbers were written down, and compared with the numerical systems in various dialects, when it at once became obvious that, although several of them could not be identified at all, the remainder were clearly connected in some way with the numerals of modern Welsh, from which latter they might at first sight appear to have been derived. In 1878, however, a paper was read on the subject by Mr. Alexander Ellis, who published in the *Transactions of the Philological Society* (in 1878, p. 316) a collection of more than fifty such sets of numerals, an analysis of which shows that these numerals came from many parts of England, and that instead of appearing most persistently in the counties bordering upon Wales, they appear with greater persistency in counties very far removed from the Welsh borders—in Surrey and Sussex, Essex and Lincolnshire, for instance. Out of the entire list, in fact, seven sets came from the northern counties, and the rest from the counties just mentioned. It was, moreover, clear (according to the view taken by Mr. Ellis himself) that the oldest of these sets of numerals might in 1878 be traced back upwards of two hundred years; and that although in modern times these numerals have come to be used by schoolboys in playing games, by nurses and mothers for amusing their babies, and by old

---

[1] The "t" at the end of "doit" (="two") is purely adventitious, and has most likely come from doubling the "t" at the beginning of the word for "three." Thus "een, doi, tree" would become "een, doi, ttree," the "t" being subsequently tacked on to the end of "doi" from the beginning of "tree" in something of the same way that the "n" in "nadder" by wrong division got tacked on to the end of the indefinite article.

Sheep-Marks and Tallies 195

women for counting the loops in their knitting, and so forth, their special association even in this degenerate stage was with the counting of sheep.

From the foregoing considerations therefore, and from their advanced state of corruption, it seemed not unreasonable to conclude that these old sheep-counting numbers are less likely to have been borrowed from Modern Welsh than to have come down to us through long ages from the time of the ancient Britons, who were left behind either as serfs, or perhaps here and there in small semi-dependent communities in various parts of the country, when the great majority of their fellows were driven into the mountains of Wales, of Cumberland (whose very name is taken from that of the ancient race), and of Cornwall. As Professor Skeat, in a letter to the author of this book, categorically puts it: "The Britons, whose traces we find in Yorkshire and elsewhere, are now restricted *in language* to Wales." Indeed, modern research seems to be coming more and more to the conclusion that the number of ancient Britons who were not expelled or forced to flee by the invaders must have been much greater than was at one time assumed. It is to be recollected, moreover, that the vast majority of the ancient Britons who remained in England must have come under Saxon masters, and must therefore have inevitably taken the lowest place in the social scale then existing. Their lot must have been little better than that of the "hewers of wood" and "drawers of water" of whom we read in the Bible, and they would tend to become the keepers of the sheep. After the Norman Conquest their former taskmasters, the Saxons, were driven to these same occupations, which they must have shared with such of the Britons as still remained. There is a trace of this stage (when the Saxons in their turn became shepherds and drovers) in the preservation of the Saxon terms, which survive to the present day in the phraseology of those professions. It must have been in this way that the Saxon shepherds acquired the sheep-counting terms from their British fellow-sufferers, viz. through sharing their hard lot.

And though the matter may be incapable of direct and conclusive proof, still having in view the distribution of these ancient methods of numeration in England, almost all being far removed from the Welsh border, it appears fair to assume that we have in these old sheep-counting scores the corrupted, broken remnants of the ancient British "score." And we may further say that this assumption appears the safer, since it would alone explain adequately the extreme state of corruption and decay into which these "scores" have fallen. For we do actually find in them just such changes as would be produced by a thousand years of possible popular interpolation, misinterpretation, and confusion. If these scores had been derived in comparatively recent times from a Modern Welsh source or sources, the disintegration would hardly have been likely to have proceeded so far as has undeniably been the case. I will now take as an example one of the new sets of the score, which still "holds" with some of the old shepherds of Lincolnshire :

No. of Sheep.	No. of Sheep.
1. Yan.	11. Yan-a-dik.
2. Tan.	12. Tan-a-dik.
3. Tethera.	13. Tethera-dik.
4. Pethera.	14. Pethera-dik.
5. Pimp.	15. Bumpit.
6. Sethera.	16. Yan-a-bumpit.
7. Lethera.	17. Tan-a-bumpit.
8. Hovera.	18. Tethera-bumpit.
9. Covera.	19. Pethera-bumpit.
10. Dik.	20. Figgit (*sic*, ?Jiggit).

Here it is probable that the forms *tethera-dik* and *pethera-dik* for thirteen and fourteen should really be written *tether-a-dik* and *pether-a-dik*, as in the case of eleven and twelve (yan-a-dik and tan-a-dik). The words for thirteen and fourteen appear to have been in fact wrongly divided, and it was no doubt in some such way as this that we come to find forms like *tethera* and *pethera*, instead of *tether* and *pether*, for three and four, the *a* representing Welsh and British *ar*. And the same remark would apply to eighteen and nineteen as well. Un-

Sheep-Marks and Tallies

fortunately there is no surviving record of any of the ancient British forms, and so we are reduced to comparing the forms of the "score" numerals with Modern Welsh, and with Old Welsh, which is only of the thirteenth century, and is technically called Middle Welsh. In the case of this particular list we should then have:

English Sheep-scores.	Middle and Modern Welsh Numerals.
4. Pether.	M.W. petuar, W. pedwar.
5. Pimp.	M.W. pimp, W. pump.
10. Dik.	M.W. dec, W. deg.
15. Bumpit.	M.W. pymthec, W. pymtheg.

In all the cases the correspondence of the modern "score" to the older forms is plain, and we should also have what is much stronger than the mere verbal correspondence between individual numerals in English and Modern Welsh—viz. an unmistakably close correspondence in the formation of the system. Thus we have:

English Sheep-scores.	Welsh Numerals.
11. Yan-a-dik.	un-ar-ddeg.
12. Tan-a-dik.	deu-ddeg.
13. Tethera-dik (*i.e.* tether-a-dik).	tri-ar-ddeg.
14. Pethera-dik (*i.e.* pether-a-dik).	pedwar-ar-ddeg.

And so forth.

When to these general and specific correspondences we add the close resemblance of fifteen, *bumpit* (English sheep-"scores"), with M.W. and Welsh *pymthec* and *pymtheg* (*pumtheg*), or even the much slighter resemblance of twenty, *figgit* (sic, ? *jiggit*), of the English sheep-score to the Welsh *igain* or *ugain* and M.W. *uncent*, it will be clear that the correspondence is more than a case of merely borrowing words. It was the *system* that was borrowed, and the English sheep-counting score and the M.W. and Welsh score are quite clearly in this respect of identical origin. Indeed, the M.W. and Welsh score of reckoning from sixteen to twenty inclusive, by recommencing the scores at fifteen and adding one, two, three, and four as units, is all but unique in the languages of the world, and this too is found in the score of the English shepherds.[1]

[1] A remarkable exception to this is recorded by the celebrated traveller and

Translated into English, the numbers of both these forms of the score, from ten upwards, would run as follows :

10. Ten.	16. One-and-fifteen.
11. One-and-ten.	17. Two-and-fifteen.
12. Two-and-ten.	18. Three-and-fifteen.
13. Three-and-ten.	19. Four-and-fifteen.
14. Four-and-ten.	20. Twenty.
15. Fifteen.	

This is about as strong an instance as could perhaps be found of reckoning by "pentads" or fives, though it is possible to find a parallel for it in the Jaloff system of numeration, once described by Mungo Park. The Jaloffs, who were negroes, on reaching fifteen, counted up to twenty by means of numbers which may be translated :

15. Ten-and-hand ;
16. Ten-and-hand-one ;
17. Ten-and-hand-two ;
18. Ten-and-hand-three ;

and so on. "Hand" in this case is, of course, used as a numeral, the equivalent of five, from the fact that the five fingers were employed by the Jaloffs in counting, as indeed they have been at some stage or other in almost every part of the world.

---

physician of Henry VIII.'s time, Andrew Borde, who in 1542 gives the Cornish numerals as follows :

1. Ouyn.		12. Dowec,	*i.e.*	two-ten.
2. Dow.		13. Tredeec,	,,	three-ten.
3. Tray.		14. Peswar-deec,	,,	four-ten.
4. Peswar.		15. Pymp-deec,	,,	five-ten.
5. Pimp.		16. Whe-deec,	,,	six-ten.
6. Whe.		17. Syth-deec,	,,	seven-ten.
7. Syth.		18. Eth-deec,	,,	eight-ten.
8. Eth.		19. Naw-deec,	,,	nine-ten.
9. Naw.		20. Igous,	,,	twenty.
10. Dec.		21. Ouyn-war-igous,	,,	one-and-twenty.
11. Unec, *i.e.* one-ten.			Etc.	

Andrew Borde adds : "No Cornish man doth number above thirty, and is named 'deec-war-negous' [*i.e.* ten-and-twenty], and when they have told thirty they do begin again, one, two, three, and so forth, and when they have recounted to 100, they say 'kans,' and if they number to 1000 they say 'myle.'"

It is very remarkable that this Cornish system of counting up to twenty, after reaching fifteen, does not follow the Welsh, but the English method. On the other hand the usual "Welsh" method of counting from sixteen to twenty is also found in Cornwall.

Sheep-Marks and Tallies 199

The peculiarity in this case, as in that of the old British and Welsh score, lies in the recommencement of the counting after fifteen is reached. The much more usual method is to reckon a complete series of teens from ten to twenty, made by adding units to ten. Our English system is different again, since our teens do not begin to run regularly in the ascending scale till thirteen is reached, the irregular forms of eleven and twelve pointing to some check in the course of development. We now come to the question of the relation in which the forms that cannot be identified with the M.W. or Welsh numbers stand to those that can be thus derived, and here we cannot do better than quote a letter from Professor Skeat, written to the Rev. W. W. Hunt of Shermanbury Rectory, which appeared in *The West Sussex Gazette*; it is now reproduced by the kind permission and courtesy of Mr. Hunt (the editor of the *Gazette*), and of Professor Skeat himself. This letter, written on June 21, 1907, ran as follows:

"The original Celtic numerals were frequently forgotten, and their places supplied by words that were more or less founded on rhyme. And sometimes the Celtic words were supplemented by English ones. Owing to the corrupt forms that thus resulted, many of the formulae are of slight philological interest or value. That the original counting was in Celtic, chiefly appears from some forms that still remain. Thus the Welsh *pump*, five, explains the Eskdale *pimp*, and the Knaresborough *pip*, and others. The Welsh *deg*, ten, explains the forms *dix, dec, dick, dik*. But *yan* (whence *yain, yaena, yah*) is only a dialectal form of the English one. And *tain, taena, tean* are merely altered forms of two, whilst the rest of the word is made to rhyme: *e.g. yain, tain, yaena, taena*; *yan, tean*; *yah, tiah*; and so on. The Welsh *pedwar*, four, has become first *peddero* (also *pethera* and *pether*) and afterwards *meddera, methera, mether*. Especially clear is the form for fifteen, when the Welsh *pymtheg*, with its variant *bymtheg* (in which the *y* is pronounced like the English *u* in pump), has given us such forms as *bumfitt, bumper*, and probably *bobtail*. How much these forms can degenerate is well shown by the Welsh *ugain* (pronounced something like *iggain*), which became *jigget* and *jigger*, and even *figget* and *ecack*."

In conclusion, it is perhaps worth while to remark that the introduction of rhyme points to a more or less conscious effort on the part of the English shepherds, who learnt these forms of the "score" from their British associates, to memorize words by no means too easy to remember without the aid of rhyme and rhythm. It is certainly to mnemonic requirements that the extension of this rhyming principle among schoolboys and children is due, the result being the production of such forms as *Eena, deena, deina, duss* (or dust), and any number of similar examples in the counting-out games of children. It was also noticed by Mr. Alexander Ellis (whose article, written in 1878, is still the standard authority on the subject) that in some of these counting-out rhymes of children another ancient and very interesting principle is illustrated, viz. that of reckoning the numbers by four at a time, in accordance with the old-world practice of counting by fours, which has left some trace even upon our own Modern English numerals, as for instance upon "eight," which, as its original termination in Old English shows, is grammatically a dual form.

WOOL HARVEST

WOOL HARVEST

When the white pinks begin to appear,
Then is the time your sheep to shear.
Old Rhyme.

SHEEP-WASHING AND SHEARING

In June wash thy sheep, where the water doth run:
And keep them from dust, but not keep them from sun.
Then shear them and spare not, at two days anende,
The sooner, the better their bodies amend.

. . . .

HUSBANDRIE FURNITURE

A skittle or skreen to rid soil from the corn,
And shearing shears ready for sheep to be shorn.

A sheep mark, a tar kettle, little or mitch,
Two pottles [1] of tar to a pottle of pitch.
THOMAS TUSSER.
From *500 Points of Good Husbandrie* (1557).

THE WASHPOOL

By RICHARD JEFFERIES, 1880

The pool is approached by a broad track—it cannot be called road—trampled into innumerable small holes by the feet of flocks of sheep, driven down here from the hills for the periodical washing. At that time the roads are full of sheep day after day, all tending in the same direction; and the little wayside inns, and those of the village which closely adjoins the washpool, find a sudden increase of custom from the shepherds. There is no written law

---
[1] A measure of two quarts.

regulating the washing, but custom has now fixed it as firmly as an Act of Parliament : each shepherd knows his day and takes his turn, and no one attempts to interfere with the monopoly of the men who throw the sheep in. The right of wash here is upheld as sternly as if it were a bulwark of the constitution.

Sometimes a landowner or a farmer, anxious to make improvements, tries to enclose the approach or to utilise the water in fertilising meadows, or in one way or another to introduce an innovation. He thinks, perhaps, that education, the spread of modern ideas, and the fact that labourers travel nowadays, have weakened the influences of tradition. He finds himself entirely mistaken ; the men assemble and throw down the fence, or fill up the new channel that has been dug ; and, the general sympathy of the parish being with them, and the interest of the sheep-farmers behind them to back them up, they always carry the day, and the old custom rules supreme.

The sheep greatly dislike water. The difficulty is to get them in ; after the dip they get out fast enough. Only if driven by a strange dog, and unable to escape on account of a wall or enclosure, will they ever rush into a pond. If a sheep gets into a brook and cannot get out—his narrow feet sink deep into the mud—should he not be speedily relieved he will die, even though his head be above water, from chill and fright. Cattle, on the other hand, love to stand in water on a warm day.

In rubbing together and struggling with the shepherds and their assistants a good deal of wool is torn from the sheep and floats down the current. This is caught by a net stretched across below, and finally comes into the possession of one or two old women of the village, who seem to have a prescriptive right to it, on payment of a small toll for beer-money. These women are on the look-out during the year for such stray scraps of wool as they can pick up from the bushes beside the roads and lanes much travelled by sheep—also from the tall thistles and briars, where they have got through a gap. This wool is more

From the painting by J. Aumonier, R.I.   *W. A. Mansell & Co., Photo.*

SHEEP-WASHING IN SUSSEX

or less stained by the weather and by particles of dust, but it answers the purpose, which is the manufacture of mops. The old-fashioned mop is still a necessary adjunct of the farm-house, and especially the dairy, which has to be constantly "swilled" out and mopped clean. With the ancient spinning-wheel they work up the wool thus gathered ; and so, even at this late day, in odd nooks and corners, the wheel may now and then be found. . . . Near the edge of the hill, just above the washpool, stands the village church. Old and grey as it is, yet the usage of the pool by the shepherds dates from still earlier days.

A SUSSEX SHEEP-SHEARING

By R. W. BLENCOWE, 1849

Solitary as the shepherds' life generally was, there was one month in the year, and that the most beautiful of all the months, that of June—the sheep-shearing month—when they met together in considerable numbers to shear the various flocks. Their work was hard ; but there was much that was enjoyable in it, for it was a season of social merriment, which contrasted strongly with the usual solitary tenor of their lives. The shearing used to be performed by companies, consisting generally of above thirty men, and most of them formerly were shepherds.[1] Each company received its distinctive name from some place within the sphere of its labours. One was called, for instance, the Brookside, another the Portslade Company ; each of them had a captain and a lieutenant placed over it, and these men, selected by the party for their trustworthy character, their superior intelligence, and their skill in the shearing art, exhibited a pleasant specimen of a good elective government. Nor were the outward

---

[1] A modern shepherd considers this statement to be incorrect. The company may have consisted of some old farm-hands, woodmen, hurdle-makers, " broom squires" (broom-makers), and such-like. Shearing-time is a slack time with woodmen. According to Mr. Fleet, writing in 1880, tailors, shoemakers, and even a stonemason were to be found among such a brotherhood. Shepherds are, as a rule, shepherds only all their lives.—[*Author's Note.*]

Wool Harvest

symbols of authority wanting, for the captain was distinguished by his gold-laced and the lieutenant by his silver-laced hat; but this distinction has now passed away. We are indebted to the Rev. John Broadwood for the following, and for other "old English songs," still sung by the peasantry of the weald of Surrey and Sussex, who collected them, after having heard them sung every Christmas from his childhood by the country-people when they went about wassailing to the neighbouring houses at that season. With the true feeling of an archaeologist he had the airs set to music exactly as they are now sung, with a view, to use his own words, to rescue them from oblivion, and to afford a specimen of old English melody. They were harmonized by Mr. Dusart, organist to the chapel-of-ease at Worthing. The stanzas, as now published, are in some degree varied from those of Mr. Broadwood, those of an old shepherd having been adopted where the variation seemed to be an improvement.

There is a springy, joyous spirit in this old Sussex song, which was sung at sheep-shearings and again at Christmas :

> Here the rosebuds in June and the violets are blowing,
> The small birds they warble from every green bough ;
>   Here's the pink and the lily,
>   And the daffadowndilly,
> To adorn and perfume the sweet meadows in June.
> 'Tis all before the plough the fat oxen go slow ;
> But the lads and the lasses to the sheep-shearing go.
>
> Our shepherds rejoice in their fine heavy fleece,
> And frisky young lambs, which their flocks do increase ;
>   Each lad takes his lass
>   All on the green grass,
>   Where the pink and the lily,
>   And the daffadowndilly, etc.
>
> Here stands our brown jug, and 'tis filled with good ale,
> Our table, our table, shall increase and not fail ;
>   We'll joke and we'll sing,
>   And dance in a ring,
>   Where the pink and the lily,
>   And the daffadowndilly, etc.

> When the sheep-shearing's over and harvest draws nigh,
> We'll prepare for the fields, our strength for to try;
>   We'll reap and we'll mow,
>   We'll plough and we'll sow;
>   Oh! the pink and the lily,
>   And the daffadowndilly, etc.

As soon as the company was formed, all the men repaired to the cottage of the captain, where a feast, which was called the "white ram," was provided for them, and on this occasion the whole plan of the campaign was discussed and arranged.

They generally got to their place of shearing about seven, and, having breakfasted, began their work. Once in the forenoon and twice in the afternoon their custom was to "light up," as they termed it; they ceased to work for a few minutes, drank their beer, sharpened their shears, and set to work again. Their dinner-hour was one, but this was not the great meal of the day, their supper being the time of real enjoyment; and when this was over they would remain for several hours in the house, smoking their pipes and singing their sheep-shearing songs, in which they were joined by the servants of the farm; and sometimes the master and mistress of the house would favour them with their presence. The following was a favourite song, and though the rhymes are anything but perfect, and here and there the metre halts, there is a rude spirit in it which will justify its being preserved:

> Come, all my jolly boys, and we'll together go
> Abroad with our masters, to shear the lamb and ewe;[1]
> All in the merry month of June, of all times of the year,
> It always comes in season the ewes and lambs to shear;
> And there we must work hard, boys, until our backs do ache,
> And our master he will bring us beer whenever we do lack.
>
> Our master he comes round to see our work is doing well,
> And he cries, "Shear them close, men, for there is but little wool."
> "O yes, good master," we reply, "we'll do well as we can."
> When our captain calls, "Shear close, boys!" to each and every
>   man;

---

[1] Pronounced *yeo*, or *yo*.—[*Author's Note.*]

And at some places still we have this story all day long,
"Close them, boys, and shear them well!" and this is all their song.

And then our noble captain doth unto our master say,
"Come, let us have our bucket of your good ale, I pray."
He turns unto our captain, and makes him this reply,
"You shall have the best of beer, I promise, presently."
Then out with the bucket pretty Betsey she doth come,
And master says, "Maid, mind and see that every man has some."

This is some of our pastime while we the sheep do shear,
And though we are such merry boys, we work hard, I declare ;
And when 'tis night, and we have done, our master is more free,
And stores us well with good strong beer and pipes and tobaccee.
So we do sit and drink, we smoke, and sing, and roar,
Till we become more merry than e'er we were before.

When all our work is done, and all our sheep are shorn,
Then home to our captain, to drink the ale that's strong.
'Tis a barrel, then, of hum cap, which we call the black ram ;
And we do sit and swagger, and swear that we are men ;
And yet before 'tis night, I'll stand you half a crown,
That if you ha'n't a special care, the ram will knock you down.

When the supper was finished, and the profits shared, they all shook hands and parted, bidding each other good-bye till another year, and each man found his way home as best he might ; on the whole, however, there was no great degree of excess. . . .

The social mirth has of late years very much abated, for since it has ceased to be the custom to shear the lambs as well as the ewes the number of men in each company has much lessened, and now the shearers frequently bring their own provisions with them and board themselves, perhaps never entering the master's house at all. Whether it be a change for the better or the worse, let others who are best acquainted with the present system decide ; but so it is.

To Mr. John Dudeney, of Lewes, the descendant of a long line of shepherds, I am indebted for all the information I have received on the subject of this paper.

210   Shepherds of Britain

From a painting by H. Singleton.   *Engraved by Cardon, 1801.*
SHEARING TIME
The old shepherd who "shears his jolly fleece" at home.

A SHEEP-SHEARING SONG

To the Tune of "Rosebuds in June"

By The Author

The song of "Rosebuds in June" is remembered by Sussex shepherds. I find the following parallel to it in

Wool Harvest

A Collection of Songs, Moral, Sentimental, Instructive, and Amusing, selected and revised by Rev. James Plumtre, M.A., vol. i., published in 1806.

THE SHEEP-SHEARING

By CHARLES JOHNSON. From *The Comedy of Country Lasses*.

When the rose is in bud, and the violets blow,
When the birds sing us love-songs on every bough,
When cowslips, and daisies, and daffodils spread,
And adorn and perfume the gay flowery mead,
    When without the plow
    Fat oxen low,
The lads and the lasses a-sheep-shearing go.

    The cleanly milk pail
    Is filled with brown ale ;
      Our table's the grass
    Where we laugh and we sing,
    And we dance in a ring,
    And every lad has his lass.

    The shepherd shears his jolly fleece,
How much richer than that which they say was in Greece.

    'Tis our cloth and our food,
    And our politic blood,
'Tis the seat which our nobles all sit on :
    'Tis a mine above ground,
    Where our treasure is found ;
'Tis the gold and the silver of Britain.

SHEEP-WASHING : AN OLD INSTITUTION NOW DECLINING

From *The Morning Post*, 1909

    Sheep-washing is one of those ancient practices which usage has sanctioned. It is not so common as it used to be, partly because of the labour involved. In the halcyon days of pastoral farming it was a poor year in which the sheep-farmer could not gather his rent from the backs of his flock. It behoved him to put his " clip " in the market in the best possible condition, and so make the highest price. But when the great foreign and colonial flock-masters began to reckon their flocks by tens of thousands the price of wool fell, and "sheep-washing" has since been gradually declining. Most farmers like a fairly deep pool in a brook in which to wash their sheep. On many farms this cannot conveniently be found, and recourse is had to the homely tub, the swim-bath, or a pond. The yolk[1] of the fleece is the only soap employed; the object of the "tubbing" being to remove the dirt which accumulates in the coat during the winter, and to wash out most of the grease which gathers in every well-conditioned skin. The fleece can then be marketed in a much finer condition than when unwashed. The difference in price between a washed and an unwashed clip is about thirty per cent.

SHEEP WASHING AND SHEARING

By John Timbs, 1867

Though it is hard and heavy work to wash and shear sheep, in the thirteenth century it was done by women, who are called "shepsters" in *The Vision of Piers Plowman* (about 1350). The sheep were washed in the mill-pond. Shearers were usually entitled to the wambelocks, or loose locks of wool under the belly of the sheep; or, as at Weston in Oxfordshire, a penny instead of the locks. The finest part of the fleece is the wool about the sheep's throat, called in Scotland the haslock, or hawselock:[2]

> A tartan plaid, spun of good hawselock woo',
> Scarlet and green he sets, the borders blew.
> Ramsay, *The Gentle Shepherd*.

Up in the north they call a sheep-shearing the clipping-time; and to come in clipping-time is to come as opportunely as at sheep-shearing, when there is always mirth and good cheer. In the middle of the seventeenth

---
[1] The greasy matter in wool.
[2] *i.e.* neck-lock ("hals" or "hawse" = neck).—[*Author's Notes.*]

Wool Harvest

century clippers always expected a joint of roasted mutton. In *The Winter's Tale* the clown ponders:

"Let me see; what am I to buy for our sheep-shearing feast? Three pound of sugar, five pound of currants, rice,—what will this sister of mine do with rice? But my father hath made her mistress of the feast, an' she lays it on.... I must have saffron to colour the warden pies; mace; dates?—none, that's out of my note; nutmegs, seven; a race or two of ginger[1]—but that I may beg; four pounds of prunes, and as many of raisins o' the sun."

The old customs of clipping-time were observed by Sir Moyle Finch, at Walton near Wetherby, in the time of Charles II., and are thus described by Henry Best:[2] "He hath usually four several keepings shorn altogether in the Hall-garth.... He hath had 49 clippers all at once, and their wage is to each man 12d. a day, and when they have done, beer and bread and cheese; the traylers[3] have 6d. a day. His tenants, the graingers,[4] are tied to come themselves, and wind the wool. They have a fat wether and a fat lamb killed, and a dinner provided for their pains. There will be usually three score or four score poor folks gathering up the locks; to oversee whom, standeth the steward and two or three of his friends or servants, with each of them a rod in his hand; there are two to carry away the wool, and weigh the roll as soon as it is wound up, and another that setteth

---

[1] Ginger in the root.   [2] *Farm Books* (Surtees), 97 (1641).

[3] Trailer. According to Wright's *Dialect Dictionary*, a "trail" is (1) a sledge without wheels, made to go down a steep hill with a load; if it had ordinary wheels a horse would be overpowered with the load. A trail is often attached behind a cart coming downhill from a stone quarry, to act as a brake and draw its own load as well.

(2) A cart with flat top and low wheels. A "trail-cart" (Gall.) is a box-cart with shafts but no wheels, mounted on a great brush of branch and twigs which stick out behind, and score the ground with ruts and scratches.

From all this it is probable that the "traylers" were those who drove the wool-sledges, or low-wheeled wool-carts.

[4] Grainger or granger, the keeper of a barn or granary; explained by the *New English Dictionary*, where this very quotation is given, as one who is in charge of a grange, a farm-bailiff; also, as here, a tenant farmer. The granger in ecclesiastical parlance was the caretaker of the garners and barns of a religious house.

Halliwell tells us that sheep-wash was a festival in the north. See *Brand's Popular Antiquities*, ed. 1841, xi. 20.

"A seed-cake at fastens, and a lusty cheese-cake at our sheep-wash."
     The Two Lancashire Lovers (1640).—[*Author's Notes.*]

it down ever as it is weighed. There is 6d. allowed to a piper for playing to the clippers all the day ; the shepherds have each his 'bell-wether's fleece'—the 'bellys' allowed to the shepherd by the old Saxon laws."

Sheep-shearing was thus celebrated in ancient times with feasting and rustic pastimes ; at present, excepting a supper at the conclusion of the sheep-shearing, we have few remains of the older custom. Nevertheless, it is interesting to revert to these pictures of pastoral life and rusticity, more especially as we find them embellished by the charms of poetry, and enlivened by a simplicity of manners which, to whatever period it may belong, is always entertaining, if not productive of better fruit.

HOLKHAM: A FAMOUS SHEEP-SHEARING FEAST

By A. M. STIRLING, 1908

Not until 1778, two years after Coke[1] had first collected the farmers together to discuss matters agricultural, did this local gathering at Holkham in Norfolk assume a definite character. . . . First, the farmers brought with them their relations and friends. These in turn brought others from a yet greater distance. Next, agriculturists from more remote parts of the kingdom wrote to ask if they might attend. Swiftly and steadily grew the fame of "Coke's Clippings," as they were called locally ; till scientists of note turned their attention to them, and men of celebrity from other countries came to England in order to be present at them ; till, year by year, they assumed greater proportions, so that they became representative of every nationality, British and foreign ; of every phase of intellect, scientific and simple; of every rank from crowned heads to petty farmers.

It was Lafayette's greatest regret that he had never witnessed a Holkham sheep-shearing. In 1818 the

---

[1] Thomas William Coke, born 1752 ; created 1st Earl of Leicester, 1837 ; died 1842. —[*Author's Note.*]

Wool Harvest 215

Emperor of Russia sent a special message to say how he wished he could be present. Among the most famous names on the page of contemporary American history are men who journeyed from the other hemisphere expressly to take part in so unique a gathering. And meanwhile the rule which had characterised the meetings in their early simplicity was never departed from; all united thus in a common interest, met on common ground; the suggestions of the simplest farmers were treated with the same respect as the conclusions of the most noted scientists; the same pains were taken in explaining to the former as to the latter the intricacies of a new system, or merits of a new implement; the same courtesy and hospitality were experienced by the most, as the least distinguished guest. . . . Never until 1821 were politics tolerated at a Holkham sheep-shearing; and then it was subsequently recognised to have been an evil omen, for that sheep-shearing proved to be the final one. Thus the "Clippings," which were always dated from 1778, extended over a period of forty-three years, until that ominous year of 1821; and during that time, it is said not a single year passed without some discovery being made, either of avoidance or adoption, and some practical benefit accruing to the human race. . . .

In 1812, Dr. Parr was present, and relates how, previous to the public gathering, he watched Coke personally working amongst his shepherds, and inspiring the men with his own remarkable energy. This confirms the account given some years before by Arthur Young, who stated how "Mr. Coke readily assists, not only his own tenants, but other neighbouring farmers. . . . He puts on his shepherd's smock and superintends the pens, to the sure improvement of the flock, for his judgment is superior and admitted. I have seen him and the late Duke of Bedford, thus accoutred, work all day, and not quit the business till the darkness forced them home to dinner."[1]

---

[1] "The Duke of Bedford afterwards emulated Coke's example, and established on his Woburn estate an agricultural meeting on the same pattern as that first instituted at Holkham."

The reputation of the Woburn estate still stands high. Mrs. Humphry Ward,

Richard Rush, while staying at Holkham for this gathering (1821), took down notes, as he says, of a "few of the things which struck me as an American and a stranger, in my visit of a week to this celebrated estate. . . . The occasion on which we were assembled," he explains, "was called the sheep-shearing; it was the 43rd anniversary of this attractive festival. The sheep-shearing conveys in itself but a limited idea of what is witnessed at Holkham. The operations embrace everything connected with agriculture in the broadest sense; such as an inspection of all the farms which make up the Holkham estate, with the modes of tillage practised on each for all varieties of crops; an exhibition of cattle, with the modes of feeding and keeping them. Ploughing matches, haymaking, a display of agricultural implements, and modes of using them; the visiting of various outbuildings, stables, and so on, best adapted to good farming; and the rearing and care of horses and stock, with much more than I am able to specify. Sheep-shearing there was, indeed, but it was only one item in this full round of practical agriculture; the whole lasted three days, occupying the morning of each, until dinner-time. The shearing of the sheep was the closing operation of the third day."[1] . . . Come what will in the future, the Holkham sheep-shearings will live in English rural annals. Long will tradition speak of them as uniting improvements in agriculture to an abundant, cordial and joyous hospitality.

THE SHEARERS' KING AND QUEEN

By WILLIAM HOWITT, June 1831

Sheep-shearing, begun last month, is generally completed this. It is one of the most picturesque operations of rural life, and from ancient times it has been regarded as a season of gladness and festivity. The simple and

writing in 1901, speaks of "This great estate of Woburn, so well and so generously managed by the Duke of Bedford."—[*Author's Note.*]

[1] *The Court of London*, by Richard Rush, ed. by his son, Benjamin Rush, 1873.

Wool Harvest

unvitiated sense of mankind taught them, in the earlier ages of society, that the bounty of Nature was to be gathered in with thankfulness and in a spirit like that of the Great Giver, a spirit of blessing and benevolence. Therefore did they join with the brightness and beauty of the summer sunshine of their grateful souls, and collect with mirth and feasting the harvests of the field, of the forest, and of the flock. Such was the custom of this country in the old-fashioned days. It was a time of merry-making; the maidens, in their best attire, waited on the shearers to receive and roll up the fleeces. A feast was made, and king and queen elected; or, according to Drayton's *Polyolbion*, the king was pre-elected by a fortunate circumstance.

> The shepherd king,
> Whose flock hath chanced that year the earliest lamb to bring,
> In his gay baldric sits at his low, grassy board,
> With flawns, curds, clouted cream, and country dainties stored;
> And while the bagpipes play, each lusty, jocund swain
> Quaffs syllabubs in cans, to all upon the plain;
> And to their country girls, whose nosegays they do wear,
> Some roundelays do sing, the rest the burden bear.[1]

Like most of our old festivities, however, this has of late years declined; yet two instances, in which it has been attempted to keep it alive on a noble scale, worthy of a country so renowned for its flocks and fleeces, will occur to the reader—those of Holkham and Woburn;[2] and in the wilds of Scotland, and the more rural parts of England, the ancient glory of sheep-shearing has not entirely departed. And indeed its picturesqueness can never depart, however the jollity may. The sheep-

---

[1] Michael Drayton alludes to the custom of decorating certain sheep with a chaplet:

> "When every ewe two lambs that yeaned hath that year
> About her new-shorn neck a chaplet then doth wear."

And also:

> "My writhen-headed ram, with posies crowned in pride,
> Fast to his crooked horns with ribbons neatly ty'd."

In *Extremesis* (1811), by Richard Fowkes, is the following entry to June 21: "Longest day—Shearing our sheep. Such dainties at village sheep-shearing, till gaping boys and men have seen the bottom of the brown jug and copious horn, and a garland of flowers on the ram's neck to grace this rural day."—[*Author's Note.*]

[2] See "A Famous Sheep-Shearing," pp. 214-216.

washing, however, which precedes the shearing, has more of rural beauty about it.

A CUMBERLAND CLIPPING

By A. W. R.

From *The Table Book of William Hone* (1827).

Letters in a recent number of *The Table Book* recalled to my mind four of the happiest years of my life spent in Cumberland amongst the beautiful lakes and mountains in the neighbourhood of Keswick, where I became acquainted with a custom which I shall attempt to describe. A few days previous to the "clipping" or shearing of the sheep they are washed at a "beck," or small river, not far from the mountain on which they are kept. The clippings that I have witnessed have generally been in St. John's Vale. Several farmers wash their sheep at the same place, and by that means greatly assist each other. The scene is most amusing. Imagine to yourself several hundred sheep scattered about in various directions, some of them enclosed in pens by the water-side; four or five men in the water rolling those about that are thrown in to them; the dames and pretty maidens supplying the "mountain dew" very plentifully to the people assembled, particularly those that have got themselves well ducked; the boys pushing each other into the river, splashing the men, and raising tremendous shouts. Add to this a fine day in June and a beautiful landscape, composed of mountains, woods, cultivated lands, and a small meandering stream; the farmers and their wives, children, and servants, with hearty faces, and as merry as summer and good cheer can make them; and I am sure, sir, that you, who are a lover of Nature in all her forms, could not wish a more delightful scene.

I will now proceed to the "clipping" itself. Early in the forenoon of the appointed day the friends and relatives of the farmer assemble at his house (for they always assist each other),[1] and after having regaled themselves with hung-beef, curds, and home-brewed ale, they proceed briskly to business. The men seat themselves on their stools with shears in their hands, and the younger part of the company supply them with sheep from the fold, which, after having been sheared, have the private mark of the farmer stamped upon them with pitch. In the meantime the lasses are fluttering about, playing numerous tricks, for which, by the by, they get paid with interest by kisses; and the housewife may be seen busy in preparing the supper, which generally comprises all that the season affords. After the "clipping" is over, and the sheep driven on to the fells [mountains], they adjourn in a body to the house; and then begins a scene of rustic merriment, which those who have not witnessed it can have no conception of. The evening is spent in drinking home-brewed ale and singing.[2] Their songs generally bear some allusion to the subject in question, and are always rural.

"RUEING" *VERSUS* CLIPPING

By THE AUTHOR

We may infer that shearing was not unknown to the Romans during their occupation of Britain, because shears similar to those used at the present time for sheep-shearing have been found in England in Roman graves. There

---

[1] Wife, make us a dinner; spare flesh, neither corn;
Make wafers and cakes, for our sheep must be shorn.
At sheep-shearing neighbours none other thing crave
But good cheer and welcome like neighbours to have.
THOMAS TUSSER (1557).

In the Lake Country it is still the custom for good cheer and welcome to be the friendly "clippers'" only payment.—[*Author's Note.*]

[2] *The Master's Good Health* is sung in Suffolk at harvest suppers, and at sheep-shearing feasts in Cumberland:

"Here's a health unto our master, the founder of the feast,
I wish with all my heart and soul in heaven he may find rest.
I hope all things may prosper that ever he takes in hand,
For we are all his servants, and all at his command.
Drink, boys, drink, and see you do not spill;
For if you do, you must drink two; it is your master's will."
[*Author's Note.*]

seems to be a dispute as to whether in earlier days the practice of plucking the wool was a cruel method, although in Young's *Tour in Ireland* (1776-1779) we read that "Lord Altamont mentioned, as descriptive of Mayo husbandry, Acts of Parliament to prevent their pulling the wool off their sheep by hand, burning their corn, and ploughing by the tail." [1]   Dr. Cowie, writing in 1871, says : " The wool is removed from the Shetland sheep not by clipping, but by the more primitive mode of 'rueing,' or tearing it out with the hand. The alleged reason for this barbarous process, which gives much pain to the poor animal, is that it ensures the fineness of the next crop of wool." Shetlanders of the present time do not consider "rueing" to be cruel. They affirm that unskilled shearers often inflict more pain by carelessness in the use of the shears than does the person that plucks the wool off. The Shetlander is tender-hearted, and, as a rule, would not take the wool until it is ready to fall off, and the particular breed of sheep that is "rued" casts its coat in the summer if it is not taken from it. Thus it seems that the method in the past was only cruel when practised on this breed of sheep at the wrong time of year, or on breeds that do not shed their wool and should be shorn (either of which case would be quite exceptional).

In Iceland and the Faroe Islands the sheep are plucked, or, as the Shetlanders would say, "rued." The report of the British consul for 1908 states that in the Faroes the plucking takes place in June, when the wool is loose, and that there are about 100,000 sheep there, which live in a half-wild state. In *An Historical and Descriptive Account of Iceland, Greenland, and Faroe Islands*, published in 1840, the writer, referring to Olafsen on the sheep of Iceland, says : "The harvest being over, the farmers employ themselves in collecting the sheep that during the summer have been wandering wild on the mountains, bringing

---

[1] Young, writing in 1776 of Farnham, near Cavan, Ireland, has : " Here they very commonly plough and harrow with the horses DRAWING BY THE TAIL, it is done every season, they insist that, take a horse tired in traces, and put him to work by the tail, he will draw better : quite fresh again. Indignant Reader, this is no jest of mine, but stubborn, barbarous truth. It is so all over Cavan."   I quote this to show with what extreme cruelty " plucking " the wool was classed.—[*Author's Note.*]

Wool Harvest

them home and killing those needed for the winter. The Icelanders do not shear this animal as in other countries, but either pull the wool off when it begins to get loose or allow it to fall spontaneously. The reason for this, according to Olafsen, is, that in cutting the wool they would also remove the long coarse hair, which is considered the principal protection from the rain, and would thus be obliged to keep them shut up during the cold season." I am told that on a rare sheep in Shetland there are those long hairs,[1] which are of a reddish colour, and that by plucking the wool these hairs are allowed to remain and form a protection against the cold.[2]

[1] Hibbert says: "The wool is short and very fine." There is very little difference since his day. Probably the reason why some have said that the Shetland sheep have hair has arisen from the fact that, when the fleece is taken off, the outer hair of the next crop is coarser than the rest, but it falls off during the year, so that nothing is taken off but the purest and finest of wool, which has become famous. — Rev. T. MATHEWSON.

[2] The word shear is thus defined by Halliwell, "to gnaw, or eat off, to tear with the teeth," and he has " roo = rough " (Devon).

THE LABOURS OF THE LOOM

By Habberton Lulham.
THE DELICIOUS DOWNS OF ALBION

> The food of wool
> Is grass or herbage soft that ever blooms
> In temperate air, in the delicious downs
> Of Albion.
> JOHN DYER, *The Fleece* (1761).

THE CARE OF WOOL
THE LABOURS OF THE LOOM

> The care of sheep, the labours of the loom,
> And arts of trade, I sing.
> JOHN DYER, *The Fleece* (1761).

OF WOOL AND WOOLLEN CLOTH
By the Very Rev. DANIEL ROCK, D.D., 1876.

SHEEP in a primitive period were bred for raiment perhaps as much as for food. At first the locks of wool torn away from the animal's back by brambles were gathered. Afterwards shearing was thought of, and followed in some countries; while in others the wool was not cut off, but plucked by the hand away from the living creature. Obtained by either method the fleeces were spun generally by women from the distaff. This very ancient daily work was followed by women among our Anglo-Saxon ancestors of all ranks of life, from the king's daughter downwards. . . .

A curious instance of the use of woollen stuff not woven but plaited, among the older stock of the Britons, was very lately brought to light while cutting through an early Celtic grave-hill or barrow in Yorkshire. The dead body had been wrapped, as was shown by the few unrotted shreds still cleaving to its bones, in a woollen shroud of coarse and loose fabric *wrought by the plaiting process without a loom.* As time passed on it brought the loom, fashioned after its simplest form, to the far west, and its use became general throughout the British Islands. The art of dyeing soon followed, and so beautiful were the tints which our Britons knew how to give to their wools that strangers wondered at and were jealous of their splendour. With regard to the bulk of the people, we learn from Dion Cassius (born A.D. 155) that the garments worn by them were of a texture wrought in a square pattern of several colours. And speaking of Boadicea, the same writer tells us that she usually had on under her cloak a motley tunic chequered all over with many colours. This garment we are fairly warranted in deeming to have been of native stuff, woven of worsted after a pattern in tints and design like one or other of the present Scotch plaids.

Our coarser native textiles in wool or in thread, or in both woven together, formed a stuff called " burel." St. Paul's in 1295 had a light blue chasuble, and Exeter Cathedral in 1277 a long pall of this texture. Burel and, in short, all the coarser kinds of work were wrought by men, sometimes in monasteries. The old Benedictine rule obliged the monks to give a certain number of hours every week-day to hand-work, either at home or in the field.

The weaving in this country of woollen cloth, as a staple branch of trade, is very old. Of the monks at Bath Abbey we are told by a late writer " that the shuttle

The Labours of the Loom 227

and the loom employed their attention (about the middle of the fourteenth century), and under their active auspices the weaving of woollen cloth (which made its appearance in England about the year 1330, and received the sanction of an Act of Parliament in 1337) was introduced, established, and brought to such perfection at Bath as rendered the city one of the most considerable in the west of England for this manufacture." Worcester cloth was so good that, by a chapter of the Benedictine Order held in 1422 at Westminster Abbey, it was forbidden to be worn by the monks and declared smart enough for military men. Norwich also wove stuffs that were in demand for costly household furniture; and Sir John Cobham, in 1394, bequeathed "a bed of Norwich stuff embroidered with butterflies." In one of the chapels at Durham Priory there were four blue cushions of Norwich work. Worsted, a town in Norfolk, by a new method of its own for the carding of the wool with combs of iron well heated, and then twisting the thread harder than usual in the spinning, enabled our weavers to produce a woollen stuff of a peculiar quality, to which the name itself of worsted was immediately given. To such a high repute did the new web grow that church vestments and domestic furniture of the choicest sorts were made out of it. Exeter Cathedral among its chasubles had several " de nigro worsted " in cloth of gold. Vestments made of worsted, variously spelt "worsett" and "woryst," are enumerated in the fabric rolls of York Minster. Elizabeth de Bohun, in 1356, bequeathed to her daughter, the Countess of Arundel, "a bed of red worsted embroidered"; and Joane, Lady Bergavenny, leaves to John of Ormond "a bed cloth of gold with leopards, with those cushions and tapettes of my best red worsted."

Irish cloth, white and red, in the reign of King John, was much used in England, and in the household expenses of Swinford, Bishop of Hereford in 1290, an item occurs of Irish cloth for lining.

English weavers knew also how to work artificially designed and well-figured webs. In the wardrobe accounts of Edward II. is this item: "To a mercer of London for a green hanging of wool wove with figures of kings and earls upon it, for the king's service in his hall on solemn feasts at London." Such "salles," as they were called in France, and "hallings," a name they went under here, were much valued abroad and in common use at home. Under the head of "Salles d'Angleterre" among the articles of costly furniture belonging to Charles V. of France in 1364, one set of hangings is thus entered: "Une salle d'Angleterre vermeille brodée d'azur, et est la bordeure à vignettes et le dedens de lyons, d'aigles et de lyepars." Here in England, Richard, Earl of Arundel, in 1392, willed to his dear wife "the hangings of the hall which was lately made in London, of blue tapestry with red roses, with the arms of my sons," etc.; and Lady Bergavenny, after bequeathing her "hullying of black, red, and green" to one friend, left to another her best stained "hall."

Flemish textiles, at least of the less ambitious kinds, such as napery and woollens, were much esteemed centuries ago; and our countryman, Matthew of Westminster, says of Flanders that, made from the material which we sent her, the wool, she sent us back precious garments. So important was the supply of wool to the Flemings in the fourteenth century, that the check given to it by the wars between England and France at that time led to a special treaty between Edward III. and the burghers of the Flemish communes under the guidance of Jacob van Artevelde.

WOOLLEN MANUFACTURE IN ENGLAND
(15TH CENTURY)

From *The Antiquary's Portfolio*, 1825

That the English in the fifteenth century had great abundance of excellent wool and were comfortably clothed, is certain from the testimony of Sir John Fortescue, who in proving that the English, who lived under a limited monarchy, were much happier than their rivals the French, who lived under a despotic government, gives the following as an example :—" The French weryn no wollyn, but if it be a pore cote, under their uttermost garment, made of grete canvas, and call it a frok, their hosyn be of like canvas, and passin not their knee ; wherefor they be gartered, and their thighs bare. Their wifs and children goen bare fote. But the English wear fine wooled cloth in all their apparel. They have also abundance of bed-covering in their houses, and all other woollen stuffe." It is probable, however, that Sir John speaks only of yeomen, substantial farmers, and artificers ; for it appears, from an Act made in 1414 for regulating the wages and clothing of servants employed in husbandry, that their dress and furniture could hardly answer the above description.

THE STORY OF THE COTSWOLD WOOL TRADE

By Francis Duckworth, 1908

In one sense the wool trade is the key to the whole of English history in the fourteenth, fifteenth, and sixteenth centuries. It determined the relations of this country with Continental Powers, particularly with Flanders, and it provided sinews for the Hundred Years' War. With regard to the Cotswolds in particular, it explains the presence of noble churches and admirable private houses. The history of the wool trade may be divided into two stages. In the first of these the whole of our wool was exported to Flanders, and since we alone produced wool, and wool was a necessity, we could impose any export duty we pleased. The second stage, or period, begins with the development of weaving in this country. More and more wool is kept in the country, until at the last export ceases altogether towards the beginning of the seventeenth century, all wool being made into cloth on English looms. The first period is practically co-extensive with the whole of the fourteenth and most of the fifteenth century. This is the period in which Chipping Campden and Northleach were built, and Cirencester began to regain something of its old importance. Their beautiful churches they owed to the piety of their richer citizens, chiefly wool merchants. This is the golden age of Cotswold, when commercial and municipal activities were so intertwined in action and co-extensive in province, that for a merchant to spend his money on enriching and beautifying his township was the rule rather than the exception.

As regards the agricultural conditions of this period, it may be noticed that the Black Death, so terrible a blow to other industries, greatly advanced sheep-farming, for seeing that sheep do not require much labour farmers gave up the plough and took to keeping sheep, till men ceased to care for anything else. " Where are our ships? What are our swords become? Our enemies bid us for a ship set a sheep," exclaims *The Libell of English Policie*. There was other evidence that the increase in the prosperity of sheep-farming was not all gain. The population soon began to recover from the decimation by plague, but the demand for labour grew smaller as more and more ploughed land was turned into pasture. "Your sheep may be now said to devour men and to unpeople not only villages but towns." [1] However that may be, the labour so liberated diverted itself into the weaving industry, which was growing at a greater pace even than the sheep-farming.

On the whole, the period from 1650 to 1750 was fairly

[1] In Bastard's *Chrestoleros* (1598) is:

"Sheep have eat up our meadows and our downs,
Our corn, our wood, whole villages and towns,
Yea, they have eat up many wealthy men,
Besides our widows and orphan children,
Besides our statutes and our iron laws,
Which they have swallowed down into their maws.
Till now I thought the proverb did but jest
Which said a black sheep was a biting beast."

Mr. Carew Hazlett makes the following comment on the above :—" Bastard merely echoes the popular panic which prevailed respecting the multiplication of sheep, and its disastrous consequences to us. In Lambeth Library is a prose tract of twelve leaves only, called, *Certayne Causes gathered together, wherein is shewed the decay of England, onely by the great multitude of shepe.*"—[*Author's Note.*]

The Labours of the Loom 231

prosperous. Much wealth was produced, even if it was not fairly distributed. Why, then, does this period leave so few traces of itself; and why do those who write of the Cotswolds slip so hurriedly by it? Because it was an unlovely age. Three things had dried up that source of beauty which flowed so abundantly in the Middle Ages; these were: the nationalisation of commerce, Puritanism, and the break-up of the manorial system. The new commerce meant a cessation of that "family" relation which existed between employer and workman; it affected quite deeply the corporate life of the villages and blighted their fine spirit of independence, so that no more money was spent in erecting and beautifying public buildings or adorning those which already existed. Puritanism helped on the gloomy work by freezing human tears and laughter, by frowning upon all who delighted in strangeness and beauty and life for its own sake, which expressed itself in the domestic architecture and the folk-songs and dances and games of the countryside. The repression of a healthy, natural activity works harm.

In 1592 Queen Elizabeth visited Sudeley Castle. Nowadays should a sovereign visit any country place there would be an address presented by the local authorities, the volunteers would turn out, and a small girl would present a bouquet of flowers, and at night perhaps there would be a display of fireworks. But that was not enough in the sixteenth century. The place had to embody its welcome, its loyalty, and its pride in some concrete form. Accordingly, at the gates of Sudeley Elizabeth was confronted by a shepherd, who spoke as follows:—

"Vouchsafe to hear a simple shepherd; shepherds and simplicity cannot part. Your Highness is come into Cotswold, an uneven country, but people that carry their thoughts level with their fortunes, low spirits but true hearts, using plain dealings, once counted a great jewel, but now beggary. These hills afford nothing but cottages, and nothing can we present to your Highness but shepherds. . . . This lock of wool, Cotswold's best fruit, and my poor gift, I offer to your Highness, in which nothing is to be esteemed but the whiteness, virginity's colour, nor to be expected but duty, shepherd's religion."

There is little grazing in Cotswold now, and the famous breed described by Drayton makes better mutton than wool.

> No brown nor sullied black the face or legs doth streak,
> Like those of Moreland, Cank, or of the Cambrian hills
> That lightly laden are; but Cotswold wisely fills
> Hers with the whitest kind; whose brows so woolly be
> As men in her fair sheep no emptiness should see.

The pastoral life will never come back to Cotswold; never again shall we hear " the happy Tityrus piping underneath his beechen bowers."

KENDAL "COTTONS"[1]

(14TH-17TH CENTURIES)

From *The Reliquary*, 1861

The second edition of Cornelius Nicholson's *Annals of Kendal* (first published 1832) is now issued. One of the most interesting chapters in the book is that of the " manufactures," including, of course, the woollen manufactures, of which Kendal can boast being the first place in the kingdom in which that business was first established. The founder of the trade here was John Kempe, a weaver from Flanders, who came over " for the purpose of practising his craft and instructing such as might desire to learn of him," and brought with him " men and servants and apprentices to the said trade." This was in 1331, in July of which year the king, Edward III., granted letters of protection to him, and offered the same " to others of the same craft and to dyers and fullers."[2] Kempe settled in Kendal,

---

[1] Not cotton goods, but woollen, called cottons, from coatings.
[2] In the *Antiquary's Portfolio* (1825) we find: "In spite of the brutal disgust of his absurd subjects, that resolute prince, bringing many Flemish and other foreign artists to settle in the island, opened a sluice to a torrent of prosperous wealth, which in the space of more than four centuries has continued to fertilize our land."—[*Author's Note.*]

The Labours of the Loom 233

and it is curious to learn that some of his descendants, bearing the same name, still reside in the locality. The trade in Kendal cottons "soon grew into repute," and became "famous everywhere." The woollen manufactures of Kendal appear to have been in highest repute, above those of other towns, about the time of Camden and Speed, in the beginning of the seventeenth century. The former writer observes: "This is a place famed for excellent clothing, and for its remarkable industry. The inhabitants carry forward an extensive trade for woollen goods, known in all parts of England." And Speed says: "This town is of great trade and resort, and for the diligent and industrious practice of making cloth, so excels the rest, that in regard thereof it carrieth a supereminent name above them, and hath great vent and traffic for her woollen cloths through all parts of England." As this was the time when Shakespeare lived, the colour Kendal Green had also achieved its popularity.

The goods were formerly carried periodically on packhorses by the makers themselves, or sent to London to be vended by the warehousemen among their customers who visited the metropolis from different parts of the kingdom. After the rise of the British colonies, North America and the West Indies, the greater part of the "Kendal cottons" were sold to merchants trading to those countries, for the clothing of the negroes and poorer planters. As the colonies increased, and slaves along with them, who were employed in the culture of tobacco in Virginia, the demand for the coarse manufacture continued to increase, till the intervention of the American War caused a total suspension of the export trade. Upon the cessation of hostilities it again revived; but our manufacturers, not able to keep pace in the improvements in machinery with those in Yorkshire, the latter interfered, and were gradually gaining advantage of Kendal, till the increase of American duties put a stop to the exportation.

CORNISH "HAIR"

By William Borlase, F.R.S., 1758

The sheep of Cornwall in ancient times were remarkably small, and their fleeces so coarse that their wool bare no better title than that of Cornish "hair," and under that name the cloth made of the wool was allowed to be exported without being subject to the customary duty paid for woollen cloth. When cultivation began to take place, and the cattle to improve in size and goodness, the Cornish had the same privilege confirmed to them by grant from Edward the Black Prince, in consideration of their paying four shillings for every hundredweight of white tin coined ; the same privilege of exporting cloth of Cornish manufacture duty free was confirmed to them by the 21st of Elizabeth. At present the eastern parts of the county, finding themselves under necessity (from the scarcity of tin) of applying themselves to tillage and pasture, from the rivers Alan and Fawy[1] eastward, have as large and fine woolled sheep as any in England, and the common people wash, card, and spin their own wool, and bring their yarn to market.

THE GREY CLOTHS OF KENT

By John Timbs, 1867

The art of making woollen cloth, which was known to the Britons, was by this time brought to perfection in England, especially in the south. "While Bradford was still the little centre of a wild hill tract in pastoral Yorkshire, the 'grey cloths of Kent' kept many a loom at work in the homesteads of Tenterden, Biddenden, and Cranbrook, and all the other little mediaeval towns that dot the weald with their carved barge-boards and richly moulded beams."[2] The distaff and the spindle, which appear to have been anciently the type, and symbol, and

---

[1] *Sic.* ? "Fowey."—[*Author's Note.*]
[2] *Saturday Review*, No. 182.

the insignia of the softer sex in nearly every age and country, were in the Saxon times still more conspicuous as the distinguishing badge of the female sex. Among our ancestors the "spear-half" and the "spindle-half" expressed the male and female line. The spear and the spindle are to this day found in their graves.[1]

"KENDAL GREEN," "COVENTRY BLUE," "LEOMINSTER ORE," "LINCOLN GREEN," AND "BRISTOL RED"

By THE AUTHOR

It is interesting to note that the poets of the fifteenth, sixteenth, and seventeenth centuries allude to the various wool manufactures. We have already quoted Cornelius Nicholson, who, in his *Annals of Kendal*, remarks on Kendal green being popular in Shakespeare's time:

> Three misgotten knaves, in Kendal green, came at my back.
> *Henry IV.*, First Part, Act II. Sc. iv.

Coventry blue was a material specially famous in Elizabeth's reign, and Coventry at one time dyed the best blue in England. "I have heard it said that the chief trade of Coventry was heretofore in making blew threde." Michael Drayton describes the dress of a shepherd and

> His breech of Cointree blue.

And in a song from Martin Peerson's *Private Music*, published 1620, the shepherdess sings:

> Is not this my shepherd swain
> Sprightly clad in lovely blue?

Was this of wool or cotton? Anyhow, there must have been a great demand for blue cloth in the past, for in Planché's *History of British Costumes* is: "Howe, the continuator of *Stow's Annals*, informs us that many years prior to the reign of Queen Mary (and therefore as early

---

[1] See "Shears on Sepulchral Slabs," page 259.

as the time of Henry VIII. at least) all the apprentices of London wore blue cloaks in summer, and in winter gowns of the same colour ; blue coats or gowns being a badge of servitude about this period." At a later period Coventry blue was also held in high reputation in America and on the Continent. At one time the village of Wonersh, in Surrey, was noted for the manufacture of blue woollen cloth, "intended for exportation to the Canary Isles."

Leominster was once a great wool centre, and Michael Drayton has :

> At Lemster, for her wool whose staple doth excel,
> And seems to over-match the golden Phrygian fell.

Again :

> Where lives the man so dull, in Britain's furthest shore,
> To whom did never sound the name of Lemster ore,
> That with the silkworm's web for smallness doth compare ?

Again :

> Her skin as soft as Lemster ore.

Herrick thus alludes to Leominster wool :

> Spongy and swelling, and far more
> Soft than finest Lemster ore.

Lincoln formerly dyed the best green in England. "The fine cloth made there was excellent, both in colour and in texture" (familiar to us from the Robin Hood ballads). Michael Drayton, in his *Ninth Eclogue*, dressed his shepherdess into a Lincoln green frock :

> She's in a frock of Lincoln green,
> The colour maid's delight.

In company with "Lincoln green" yet another famous colour, "Bristol red," was referred to in the fifteenth century by Skelton, the Poet Laureate of Henry VII., as being worn by the poorer classes, typified in this particular context by Eleanor Rumming.[1]

[1] MS. Harl. Lib. 7333.

THE GREAT FAIR OF STOURBRIDGE
(14TH CENTURY AND AFTER)

By JAMES E. THOROLD ROGERS, 1882

Of these fairs, the most important for the whole east and south of England were: the great fair at Stourbridge, held under the authority and for the profit of the Corporation and city of Cambridge; the cattle fair at Abingdon; and a fair at Winchester, chiefly held for the sale of produce and cloth. But the Stourbridge fair was by far the most considerable, and was commenced and concluded with great solemnity. It was proclaimed on the 4th of September, opened on the 18th, and continued for three weeks. It is said, that the origin of the fair was in the casual establishment of a mart for the sale of Kendal cloth. . . . The temporary buildings erected for the purposes of the fair were, by custom, commenced on the 24th of August; the builders were allowed to destroy the corn grown on the spot if it were not cleared before that time; and on the other hand, the owner of the soil was empowered to destroy the booths on Michaelmas day if they were not removed before that time.

The space occupied by the fair, which was about half a mile square, was divided out into streets,[1] in each of which some special trade was carried on, some of the principal being those of ironmongery, cloth, wool, leather, and books; as well as, in the course of time, every conceivable commodity that could be made and sold. The port of Lynn, and the rivers Ouse and Cam, were the means by which water carriage was made available for goods. . . .

The concourse must have been a singular medley. Besides the people who poured forth from the great towns —from London, Norwich, Colchester, Oxford, places in the beginning of the fourteenth century of great comparative importance . . . there were, beyond doubt, the repre-

---

[1] Actually called "rows." Traces of these old rows still exist on the very spot where the fair was once held. Thus "Garlic Row," now the name of a small mean street, has got its name from the "row" where garlic was sold in the Stourbridge Fair. —[*Author's Note.*]

sentatives of many nations collected together to this great mart of mediæval commerce. . . .

Blakeney, and Colchester, and Lynn, and perhaps Norwich, were filled with foreign vessels, and busy with the transit of various produce, and eastern England grew rich under this confluence of trade. . . .

To this great fair . . . came the wool-packs which then formed the riches of England, and were the envy of outer nations.

"SHEPHERD'S PLAID"

By J. R. Planché, 1834

Mr. Logan informs us that woollen cloths "were first woven one colour, or an intermixture of the natural black and white, so often seen in Scotland to the present day." And we may add that it will be recognised by our readers as the stuff lately rendered fashionable for trousers, under the name of "Shepherd's plaid." The introduction of several colours dates from the earliest period of its manufacture; and it is asserted, both in Ireland and in Scotland, that the rank of the wearer was indicated by the number of colours in his dress, which were limited by law to seven for a king or chief, and four for the inferior nobility, while, as we have already quoted from Heron, it was "made of one or two colours" (that is to say plain, or merely chequered with another colour) "for the poor." [1]

SHETLAND WOOL

By Robert Cowie, M.D., 1871

The great merit of Shetland wool is its fineness, for it is too soft to be very durable; it is capable of being spun into threads more delicate than those that form

---

[1] A friend of mine, writing from Kingussie, Scotland (1909), says: "My landlord, who is an old 'wool miller,' weaves the shepherd's plaid. It is of black and white check—four threads black, four white, or, if a larger check is wanted, eight black and eight white, but always in fours. I often see shepherds in bonnet and plaid."—[*Author' Note.*]

the finest lace, and with such, worsted stockings have been made which could be drawn through a lady's finger ring! The worsted is generally preserved in its natural colour, but it is sometimes dyed.[1] In former times various

"SHEPHERD'S PLAID" (1851)

native dyes were used for this purpose. They are still employed, especially in the more remote districts, but they have, to a great extent, been superseded by indigo and cochineal, imported from the south. The variegated

[1] The old Norn names by which the various colours of the wool of the Shetland sheep were known are still remembered.—[*Author's Note.*]

and fantastic hues which characterise such articles as the Fair Isle hosiery, and the more commonplace hearth-rugs, are obtained by means of these native dyes, most of which are lichens. The *Lichen tartareus* yields a reddish purple, the *L. omphaloides* a blackish purple, the *L. saxatilis* (called old man) a yellowish or reddish brown, and the *L. parietinus* (termed by the Shetlanders "scriota") an orange dye. Yellow is obtained from a collection of plants of that colour, and black from peat-moss impregnated with bog iron ore. . . .

It would appear that the rearing of sheep in Shetland was attended with far greater success in ancient times than within the last two or three generations. Larger quantities of wool were devoted to the preparation of a kind of coarse cloth termed "wadmel,"[1] the manufacture of which was for ages one of the chief industries of the country. It was in this fabric that the Shetland Udallers were in the habit of paying their "scat"[2] or land-tax to the ancient kings of Denmark. The native dyes above mentioned are said to have been extensively used in colouring "wadmel." Its manufacture still existed in the beginning of the last century, but was rapidly on the decline. At the present day a considerable quantity of "claith," or flannel cloth, is made on hand-looms. The "Shetland tweeds" of the southern markets are very soft and elastic, and much prized by sportsmen for shooting suits. Since the islands came into such close commercial intercourse with the Scotch ports, cotton underclothing has to a great extent superseded the old woollen home-spun garments, and the health of the Shetlanders is said to have declined accordingly. The skins of the Shetland sheep are either dried with the wool adhering, and used as mats, or tanned with a native plant, the *Tormentilla*

---

[1] Also "wadmol."
[2] Tribute or tax. Hence the expression scot-free, *i.e.* free of tax. (*Scat*, Danish.) The following record of tithe-paying, from Edmondston's *Zetland* (1809), is interesting : "Observe, that in the year of our Lord 1328, the 25th day of July, did Giafaldr Ivarson of Hialtland pay to the Reverend Lord Audfin, the Lord Bishop of Bergen, and Swein Sigurdson, Comptroller of the King's household, the tenths due to the Pope, viz. 22 cwt. of wool, less 16 pounds, according to the standard of Hialtland, being 36 span Hialtland weight of wool."—Translated from the original Danish, in the *Bibliotheca Britannica*. Antiquarian Society.—[*Author's Note.*]

officinalis, and made into waterproof clothes for the men at the fishing-grounds, or for the women when engaged in their more severe labours in the hills and fields.

WOOLLEN CLOTHS OF IRELAND

From *The Antiquary's Portfolio*, 1825

The very necessary art of making woollen cloth (introduced, or at least highly improved in England by colonies of Flemings) seems to have flourished more in the eleventh and twelfth centuries than in those immediately succeeding. This may be reasonably accounted for by the civil wars which desolated the island and ruined every species of commerce and manufacture under Stephen, John, and Henry III. And here, in justice to our Sister Island, we must not omit to bring forward the testimony of an Italian poet and traveller, Fazio degli Uberti, who, in his *Ditta Mondi*, thus records the serges or "says" of Ireland at the beginning of the fourteenth century:

> Similimente passamo in Irlanda,
> La qual fra noi è digna di fama
> Per le nobile saie che ci manda.

Which is imitated as follows:

> To Ireland then our sails we raise;
> Ireland, which merits well our praise,
> By sending us its noble says.

The *Dictionary della Crusca* speaks of Irish "says"; and Madox and Rymer are not silent concerning the friezes and other woollen manufactures of Ireland in the time of Henry III. and Richard II. These circumstances give to the Irish the priority of a steady woollen manufacture.[1]

A FAMOUS COAT-MAKING WAGER

By WALTER MONEY, F.S.A.

A picture painted by Luke Clint, and drawn on stone by J. W. Giles, has for its subject the winning of a wager

[1] *Transactions of the Royal Irish Academy.*

for 1000 guineas by the making of a coat from freshly shorn wool between sunrise and sunset, in 1811. The contemporary account is as follows :

"To Robert Throckmorton Esq$^{re.}$ Buckland House Farringdon.

"This print representing the beginning progress and completion of an extraordinary undertaking to prove the possibility of wool being manufactured into cloth and made into a coat between sunrise and sunset, which was successfully accomplished on Tuesday 25th June 1811. Is respectfully dedicated by his obliged and humble servant John Williams, Land Steward to the late Sir John Throckmorton.

"On the day above stated at 5 o'clock in the morning Sir John Throckmorton a Berkshire baronet presented 2 Southdown sheep to Mr. Coxeter of Greenham Mills near Newbury Berkshire. The sheep were immediately shorn, the wool sorted and spun, the yarn spool'd, warp'd, loom'd, and wove. The cloth burr'd, mill'd, row'd, dy'd, dry'd, sheared and pressed. The cloth having thus been made in 11 hours was put into the hands of the tailor's at 4 o'clock in the afternoon, who completed the coat at 20 minutes past six. Mr. Coxeter then presented the coat to Sir John Throckmorton who appeared with it the same evening at the Pelican Inn, Speenhamland. The cloth was a hunting kersey of the admired dark Wellington colour. The sheep were roasted whole and distributed to the public with 120 gallons of strong beer. It was supposed that upwards of 5000 people were assembled to witness this singular unprecedented performance which was completed in the space of 13 hours and 20 minutes. Sir John and about 40 gentlemen sat down to a dinner provided by Mr. Coxeter and spent the evening with the utmost satisfaction at the success of their undertaking."

SHEPHERDS' GARB

A SHEPHERD OF 1836

With "bottle," wearing buskins, and sheepskin cloak strapped plaidwise across the shoulder.

SHEPHERDS' GARB

SHEPHERDS' DRESS, PAST AND PRESENT

By The Author

In the most ancient period the shepherds of Britain wore a sheepskin cloak, which was fastened, as among the ancient Germans, with a long thorn. In the time of the Anglo-Saxons a tunic was worn, which has survived down to the present under various forms, as the shepherd's smock or "hamp," a term which still lingers on in some out-of-the-way parts of the country. The Rev. J. C. Atkinson, in *Forty Years in a Moorland Parish*[1] (1892), writes: "There was a time when the hamp was the English peasant's only garment; at all events, mainly or generally so. For it might sometimes be worn over some underclothing, but that was not the rule. The hamp was a garment of the smock-frock type, gathered in somewhat about the middle, and coming some little way below the knee. The mention in *Piers Plowman*[2] of the 'hatere'[3]

---

[1] Messrs. Macmillan & Co. (by permission).
[2] Written, it is supposed, about the year 1350.
[3] Hatere = O.E. dress, clothing, attire.

worn by the labouring man in his day, serves to give a fairly vivid idea of the *attire* of the working man of the fourteenth century, and that attire was the 'hamp' of our northern parts; for the word seems to be clearly Old Danish in form and origin."[1]

F. W. Fairholt gives us the following interesting picture of the costume of a shepherd on holiday occasions in the fourteenth century. It is taken from *A Tale of King Edward and the Shepherd*, which was published in Hartshorne's *Metrical Tales*.[2]

> On morow, when he should to court goo,
> In russet clothing he tyret hym tho,[3]
> In kyrtil and in curtpye,[4]
> And a blak furred hode
> That wel fast to his cheke stode,
> The typet myght not wrye.[5]
> The mytans clutt forgate the nozt,
> The slyng even ys not out of his thozt,
> Wherewith he wrouzt maystre.[6]

In *The Antiquary's Portfolio* the dress of the labourers and "common people" in the fifteenth century is described as simple and well contrived, "consisting of shoes, hose made of cloth, breeches, a jacket, and coat buttoned and fastened about the body by a belt. They covered their heads with bonnets of cloth. As they could not afford to follow the caprices of fashion, the dress of both sexes continued nearly the same for several ages."[7]

Again, in Richard Hill's *Commonplace Book* (**MS.**), a poem of eleven verses (written late fifteenth century or before 1504, though the poems, being a collection, may themselves have a much earlier origin), we meet with the following description of a shepherd's dress :—

---

[1] "Hamp" = Dan. "hampe," a peasant's frock.—*English Dialect Dictionary*, Wright.
[2] Published 1829; printed chiefly from original sources.
[3] He dressed him then.
[4] Orig. "surstbye" (*sic*): probably "courtpye," a short outer garment or mantle.
[5] His hood was so well secured that the tippet conld not go awry.
[6] His mittens, and the sling, in the use of which he was famous, he also carried with him.
[7] See picture of shepherd with pipe, page 281.

Shepherds' Garb

Verse I

The shepherd upon a hill he sat,
He had on him his tabard [1] and his hat,
His tar-box, his pipe, and his flagat,
His name was called Jolly, Jolly, Wat!

It is remarkable that a few lines later, in the same poem, we meet with the lines :

He put his hand under his hood,
He saw a star as red as blood.

This passage may perhaps indicate better than any other the transition period when "hats" superseded the ancient traditional hood. Yet, again, we have Michael Drayton's fine description of a shepherd's dress in the sixteenth century :

The shepherd wore a *sheep-grey* cloak,
Which was of the finest lock
That could be cut with sheer.[2]
His mittens [3] were of buazon's [4] skin,
His cockers [5] were of cordiwin,[6]
His hood [7] of miniver.
His aul and lingel [8] in a thong,
His tar-box on his broad belt hung,
His breech of Cointree blue.[9]

Note sheep-grey, not sheep-white, perhaps as signifying the prevailing tint at that time. In fact, poets often wrote of homespun grey. See "Merry Dick" by Dibdin (1771-1841) and "The Happy Shepherd" (anon.), in Plumptre's *Songs*, published 1806.

"In 1571 felt hats were not made in England, as a statute was then enacted which ordered an English woollen cap

---

[1] A fourteenth-century garment ; "a species of mantle which covered the front of the body and the back, but was open at the sides from the shoulders downwards." Strutt describes it as a light vestment or sleeveless coat.—[*Author's Note.*]
[2] Shears.   [3] Gloves.   [4] Badger.   [5] High boots.   [6] Spanish leather.
[7] Still the hood in sixteenth century ; this is long after the introduction of hats.
[8] Shoemakers' thread.
[9] Coventry blue.
Compare Fletcher's *Faithful Shepherdess* (1590) :

Every shepherd boy
Puts on his lusty green, with gaudy hood,
And hanging scrip of finest cordevan.
[*Author's Note.*]

to be worn in preference, by every person above the age of seven, on pain of forfeiting three shillings and fourpence, lords, gentlewomen, etc., excepted."[1] William Hone in his *Year Book* tells us that in the reign of Edward IV. a law was made respecting "apparel" and reads thus : "No servant of husbandrie, or common labourer, shall weare in their clothing any cloth whereof the broad yard shall pass the price of two shillings ; nor shall suffer their wives to weare kerchiefs whose price exceedeth twenty pence." And in Elizabeth's reign : " All persons above the age of seven years shall wear upon Sabbaths and holidays, upon their heads, a cap of wool, knit, thicked, and dressed in England, upon forfeit, for every day not wearing, 3s. and 4d."

Aubrey remembered that " before the Civil War many of them made straw hats, which was then left off; and the shepherdesses of late years do begin to work point, whereas before they did only knit stockings."

Again, in respect to the seventeenth century, we read in Chambers's *Book of Days* : "Since 1671 " (when Aubrey wrote) "shepherds are grown so luxurious as to neglect their ancient warm and useful fashion, and go *à la mode*. . . ."

The smock so much worn by shepherds was sometimes of flax and sometimes of hemp, which latter plant grew in Northern Europe, whence first came, no doubt, the garment of the type of the "hempen homespuns" of which Shakespeare speaks in *A Midsummer Night's Dream* : "what hempen homespuns have we swaggering here ?" Frederic Shoberl in 1813[2] alludes to the Suffolk hemp as "superior in strength and quality to that of Russia. The cloths woven from it are of various degrees of fineness and breadth, from 10d. a yard half ell wide to 4s. to 4s. 6d. ell wide. Low-priced hemps are a general wear for servants, husbandmen, and labouring manufacturers, those from 18d. to 2s. a yard for farmers and tradesmen. The tract in which hemp is chiefly found extends from Eye to Beccles, and is about 10 miles in

---

[1] *Antiquary's Portfolio.*   [2] *The Beauties of England and Wales.*

breadth. It is cultivated both by farmers and cottagers, though it is very rare to see more than 5 or 6 acres in the hands of one person."

Coming to the present times, a few old men still

BOB PENNICOTT. IN SHORT SMOCK (STILL WORN IN LIN-
COLNSHIRE), CARRYING "BOTTLE," BELL, AND CROOK.
"JACK" IN ATTENDANCE

wear the smock in Shropshire, usually made of "hempen homespun." A really old specimen of these smocks in perfect condition is said to be worth about £10, an imperfect one from perhaps £3. They were worn with the top-hat on Sundays, and the pattern on the front varied according to the part of the country that the smock came from, the Shropshire pattern having a silk thread at the top, above the "honeycomb" pattern, "as

SHEPHERD SMITH OF WASHINGTON, NEAR CHICHESTER, IN SMOCK AND "CHUMMEY"

long as the Shropshire people's noses" (for they are celebrated for their long noses in Shropshire)!

Mrs. Wild, aged eighty-three, who lives on the Goodwood estate, and whose brother was a shepherd on the Downs for over seventy years, tells me that the

Shepherds' Garb

shepherds wore blue smocks when she was young (but drab and grey are said to have been most usual in Sussex), cloth cloaks in cold weather, and for wet weather cloaks made of unbleached calico, brushed over with a preparation of boiled oil and lamp-black—this rendered them waterproof and "gave a colour." The men often cut them out themselves, and their wives made them up, adding large outside pockets. High leggings reaching well above the knee were to match. A shepherd would often "waterproof" enough unbleached calico to make a cloak and leggings for himself and one or two neighbouring shepherds. The process required care; if the mixture was applied too thickly it cracked in the sunshine. The warm winter cloaks were generally discarded military cloaks. Their straw hats were made by both men and women, the straw being first put into boxes containing brimstone to bleach it—each straw split into four, then plaited and stitched into shape to suit the fancies of the wearer. Their soft felt winter hats they still call "chummeys."

SMOCKS AND THEIR WEARERS

By WILLIAM HOWITT, 1831

In the counties round London, eastward, and westward through Berkshire, Hampshire, Wiltshire, etc., the English peasant, shepherd, and drover is the white-smocked man of the London prints. In Hertfordshire, and in that direction, he sports his olive-green smock. In the Midland counties, especially Leicestershire, Derby, Nottingham, Warwick, and Staffordshire, he dons a blue smock called the Newark frock, which is finely gathered in a square piece of puckerment on the back and breast, on the shoulders, and at the wrists; is adorned also in those parts with flourishes of white thread, and as invariably has a little white heart stitched in at the bottom of the slit at the neck. A man would not think himself a man if he had not one of these smocks, which are the first things that he sees at a market or a fair, hung aloft at the end of the slop-vendor's stall, on a crossed pole, and waving about like a scarecrow in the wind. Under this he generally wears a coarse blue jacket, a red or yellow shag-waistcoat, stout blue worsted stockings, tall laced ankle boots, and corduroy breeches or trousers. A red handkerchief round his neck is his delight, with two good long ends dangling in front. In many other parts of the country he wears no smock at all, but a corduroy or fustian jacket, with capacious pockets, and buttons of a giant size.[1]

[1] Mr. J. K. Fowler tells us in *Recollections of a Country Life* (1894; Messrs. Longmans, Green & Co.) that when a smock is worn at meal-times, as it stretches across the knees, a sort of bag is formed, and at the termination of the meal the crumbs and small portions of cheese, etc., collected there are caught up in the hand and forthwith thrown into the mouth. Old Sussex shepherds corroborate this.—[*Author's Note*.]

SHEPHERDS' ARTS, IMPLEMENTS, AND CRAFTS

SHEPHERDS' ARTS, IMPLEMENTS, AND CRAFTS

THE PASTORAL CROOK

By Richard Jefferies, 1887

The shepherd was very ready and pleased to show his crook, which, however, was not so symmetrical in shape as those which are represented upon canvas. Nor was the handle straight; it was a rough stick—the first, evidently, that had come to hand. As there were no hedges or copses near his walks, he had to be content with this bent wand till he could get a better. The iron crook itself, he said, was made by a blacksmith in a village below. A good crook was often made from the barrel of an old single-barrel gun, such as in their decadence are turned over to the birdkeepers. About a foot of the barrel being sawn off at the muzzle end, there was a tube at once to fit the staff into, while the crook was formed by hammering the tough metal into a curve upon the anvil. So the gun—the very symbol of destruction—was beaten into the pastoral crook, the implement and emblem of peace. These crooks of village workmanship are now subject to competition from the numbers offered for sale at the shops at the market towns, where scores of them are hung up on show, all exactly alike, made to pattern, as if stamped out by machinery. Each village-made crook has an individuality, that of the blacksmith—somewhat crude, perhaps, but distinctive—the hand shown in the iron.

From a painting by Alfred Parsons, A.R.A. THE WOOD-CARVER *W. A. Mansell and Co., Photo.*

"I recollect when the shepherds in **Berkshire** and **Wiltshire** used to while away the time by carving sticks with all **sorts of curious devices**."—WALTER MONEY, F.S.A.

SHEPHERDS' CROOKS

By E. V. Lucas, 1904

Pyecombe, in Sussex, has lost its ancient fame as the home of the best shepherds' crooks, but the Pyecombe crook for many years was unapproached. The industry has left Sussex; crooks are now made in the north of England, and sold over shop counters. I say "industry" wrongly, for what was truly an industry for a Pyecombe blacksmith is a mere detail in an iron factory, since the

PYECOMBE HOOK

number of shepherds does not increase, and one crook will serve a lifetime and more. An old shepherd at Pyecombe, talking confidentially on the subject of crooks, complained that the new weapon as sold at Lewes, although nominally on the Pyecombe pattern, is "a numb thing." The chief reason he gave was that the maker was out of touch with the man who was to use it. His own crook (like that of Richard Jefferies' shepherd friend) had been fashioned from the barrel of an old muzzle-loader. The present generation, he said, is forgetting how to make everything; why, he had neighbours, smart young fellows too, who could not even make their own clothes!

HOOK AND CROOK

By The Author

The Pyecombe hook or "crook" is not known in the Sheffield district. In fact, the demand for crooks is but small. The manufacturers say that the shepherds prefer those made by local blacksmiths, who can best carry out their "whims." Orders for them also are frequently executed by the smiths attached to good-class ironmongers. This, at least, is the case in the eastern counties, in Warwickshire, and in Dorsetshire. Some of the village blacksmiths in Sussex still make crooks to order. At West Dean the demand is for about three in the course of a year. These simple implements take a skilled smith about two hours to fashion; but every shepherd has his own fancies as to exact shape, and in Sussex they always decide on "the Pyecombe." In Chichester two of the ironmongers make them, and each sells about six a year. These, also, are hand-made, and of the same favourite shape, which the Southdown shepherds prefer for their small sheep. Now and again an ingenious shepherd would amuse himself by carving the handle of his crook, and Michael Drayton (in his *Pastorals, Tenth Eclogue*) tells us of one of these carved handles :

> When on an old tree, under which ere now
> He many a merry roundelay had sung,
> Upon a leafless canker-eaten bough
> His well-tun'd bagpipe carelessly he hung :
> And by the same, his sheep-hook, once of price,
> That had been carved with many a rare device.

THE CARVING OF CROOKS

Shepherds' crooks were exhibited in London in 1901 in the Rooms of the Society of Antiquaries at Burlington House, and thus described in the catalogue :[1]

"Three shepherds' crooks, varying in length, the longest quite plain; the medium one has the name of 'Seaforth' stamped on the crook. The shortest of the three is elaborately carved; rose leaves and tendrils, the leaf and flower of the thistle, and three snakes, one of them with its tail commencing from the end of the stick, and finishing in the head at the base of the crook; the rough appearance of the thistle is simulated, and a garter motto bearing the inscription, 'TIR NAN Beann,' surrounds

[1] See *International Folk-Lore Congress*, page 434. (By permission of Folk-Lore Society.)

Shepherds' Arts, Implements, & Crafts

a naked arm holding a sword. Surmounting this is an antlered stag's head as crest. The inscription should be 'Tir nam Beann,' *i.e.*, in Gaelic, 'The land of the mountains,' or 'The land of Bens.' It is not the motto of any clan, but simply a Highland and Celtic sentiment. The stag's head and antlers is the crest of the clan M'Kenzie, Seaforth being the name of the chief of the clan. Their motto, however, is *Caberfeidh*, or 'Deer's Antlers,' and it has been suggested by a Highland gentleman that, as this motto is absent, the carving on the crook is simply a fanciful and artistic device. Highland laddies carve crooks and sticks in the winter for the annual summer market."[1]

Two hundred years ago shepherds used to have a hollow piece of iron or horn at the other end of their crooks, by which they took up stones or bits of turf to throw at their flocks, as noted in the page on slinging.

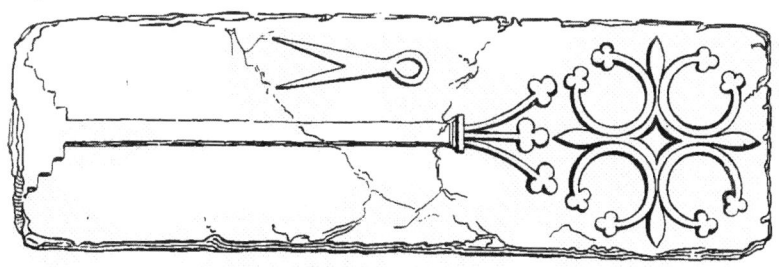

OF SHEPHERDS' SHEARS

By The Author

In *Half-Hours with English Antiquities* (1880) Mr. Llewellynn Jewitt, F.S.A., remarked as follows :—" One of the commonest devices upon early sepulchral slabs is the shears. This device has given rise to much controversy, and to considerable difference of opinion; some authorities maintaining that it denotes the deceased to have been a

---

[1] Shepherds in Berkshire used to carve sticks or staffs ; see page 256.

wool-stapler or clothier, others that he was a merchant of the staple, and others that it symbolised a female. It is sometimes found in connexion with a sword, or with a key, or two keys; but usually alone at the side of the shaft of the cross. At Aycliffe is a double slab. One half has a cross with a sword on one side of the shaft, and on the other a pair of pincers and a T-square. The other has on one side of the shaft a pair of shears, and on the other a key. There are also three crosses patée on different parts of the slab. The shears are usually of the old shape still retained in our sheep-shears of the present day, occasionally varied in details; but instances occur, as at Bakewell, where the ends of the blades are broad instead of pointed."

It is possible that, by collecting a sufficient number of examples, this point might be elucidated. Meanwhile all that can be said is that these emblems given above are clearly symbolical. And as a sword, even if combined with the shears, must be a man's emblem, and the pincers and T-square are the emblems of a carpenter's trade, it is possible that the shears or shears and key may be the emblems of a woman. With regard to the possibility of the shears being a woman's emblem, it is important to recall that women were once employed as sheep-shearers, as we learn from *Piers Plowman*, and were called "shepsters," and that this practice continued in parts of Scotland to within living memory. At the present day, however, the practice seems nearly obsolete, for an authority on these matters, writing a few days ago from Carstairs, says: "I have only known one female shearer in Scotland, a shepherd's daughter, and a regular 'tomboy.'"

SHEPHERDS' SLINGS

By Joseph Strutt, 1801

The art of slinging, or casting of stones with a sling, is of high antiquity, and probably antecedent to that of archery, though not so generally known, nor so generally

Shepherds' Arts, Implements, & Crafts

practised. . . . It was an instrument much used by the shepherds in ancient times, to protect their flocks from the attacks of ferocious animals ; if so, we shall not wonder that David, who kept his father's sheep, was so expert in the management of this weapon. . . . I remember in my youth to have seen several persons expert in slinging of stones, which they performed with thongs of leather, or, wanting those, with garters ; and sometimes they used a stick of ash or hazel, a yard or better in length, and about an inch in diameter ; it was split at the top so as to make an opening wide enough to receive the stone, which was confined by the reaction of the stick on both sides, but not strong enough to resist the impulse of the slinger.[1] It required much practice to handle this instrument with any degree of certainty, for if the stone in the act of throwing quitted the sling either sooner or later than it ought to do, the desired effect was sure to fail. Those who could use it properly cast stones to a considerable distance, and with much precision. In the present day the use of these engines seems to be totally discontinued. Barclay, in his *Eclogues* (sixteenth century), has made a shepherd boast of his skill at archery, to which he adds :

> I can dance the raye,[2] I can both pipe and sing,
> If I were merry ; I can both hurle and sling.

WHEATEAR[3] TRAPPING BY SHEPHERDS

By Arthur Beckett, 1909

Sauntering one summer afternoon along a low slope of the South Downs, my attention was attracted by a flutter-

---

[1] F. W. Fairholt remarks that "the sling appears to have been a leathern bag fixed to the end of a staff and wielded with both hands. They were much used by shepherds." This, however, was not the only method of slinging formerly employed in England, for Evelyn (1620-1706) notes that for slings the shepherds had in his time a hollow iron or piece of horn, not unlike a shoeing horn, fastened to the other end of the crosier, by which they took up stones and kept their flocks in order.—[*Author's Note.*]

[2] Strutt remarks as follows of the ray : "The ray (or raye, as it is written by Chaucer) appears to have been a rustic dance, and probably the same as that now called the hay, where they lay hold of hands and dance round in a ring. A dance of this kind occurs several times in the Bodleian MS., date A.D. 1344."—[*Author's Note.*]

[3] *Saxicola œnanthe.*

ing under an upturned clod of turf. I stepped to the spot and saw a wheatear entangled in a trap by a horse-hair noose around its neck. The little thing was terribly alarmed at my presence, but pulling a penknife from my pocket, I freed it with a touch of the blade, and in a flash it had flown out of sight. Then, in accordance with the custom, I settled my account with the proprietor of the trap by placing some pence under the upturned sod of turf.

The South Down country is the summer home of the timid wheatear, a beautiful bird, with black wings and grey and white body plumage. Its sweet low note of "far-far" and "titreu-titreu" marks it as the softest singer among the merry and more musical songsters of the Downs. It is an excellent mimic, and the note of the male bird, particularly, is pretty. . . . For hundreds of years the wheatear has been highly esteemed as a table delicacy, earning from *gourmets* the title of the "English ortolan," the flavour of which it is said much to resemble. . . . The trade in wheatears was formerly a staple industry among the shepherds of the South Downs. . . . The manner of trapping the bird was peculiar. A series of T-shaped shallow trenches, each about a foot long, was dug on the slope of a hill. The turf from each excavation was removed, and a horsehair spring set at the inner end. The sod was then replaced on the top of the trap, grass downwards, a small opening being left at the lower extremity of the T, by which the bird might enter. The wheatear owed its capture in these traps to its excessive timidity. Its habit is to skim the ground in flight, and at the least alarm, such as the shadow cast by a cloud, it seeks shelter in the nearest hole or cranny. The shepherd's knowledge of this characteristic enabled him to invent the peculiar form of trap to which I have referred, and in the month of July, when the birds are most plentiful on the South Downs, and the shadows of the clouds chase one another across the uplands, a shepherd would often take birds from the snares three or four times a day. Later in the year—during the months of August and September—a single shepherd has been known, by means of these traps or "coops," as they were commonly called, to capture between eighty and ninety dozen wheatears in a day; and when the birds have assembled previous to migration certain parts of the Downs have been honeycombed with traps.

Pennant states that when the trade in these birds was at its best, one thousand eight hundred and forty dozen wheatears were annually ensnared by the shepherds in the Eastbourne district alone. In 1842 sixty dozen were sent to London in one day by the Eastbourne coach. . . .

Dudeney sold his wheatears at 2s. 6d. or 3s. a dozen, though the standard price given by poulterers was 1s. 6d. In 1665 the Rev. Giles Moore, a Sussex diarist, records that he bought two dozen at Lewes for 1s. But, as is usually the case, there is no doubt that the prices were governed by the laws of demand and supply. . . .

A favourite method of cooking wheatears was to wrap each bird in vine-leaves and roast it. . . . Sometimes those who bought wheatears from the shepherds would visit the Downs and take the birds themselves from the springs, leaving their market price in the traps to be collected by the shepherds later in the day. Reading recently in *The Favourite Village*, by James Hurdis, an almost forgotten Sussex poet, I came across the following references to this custom:

> When the fevered cloud of August day
> Flits through the blue expanse,
> The timorous wheatear, fearful of the shade,
> Trips to the hostile shelter of the clod,
> And where she sought protection finds a snare.

> Seized by the springe
> She flutters for lost liberty in vain,
> A costly morsel, destined for the board
> Of well-fed luxury, if no kind friend,
> No gentle passenger the noose dissolve,
> And give her to her free-born wing again.

> To the feathery captive give release,
> The pence of ransom placing in its stead.[1]

---

[1] An old shepherd tells me that 3d. was the proper sum. The birds were sometimes stolen from the traps, but not often.—[*Author's Note.*]

And so, remembering Hurdis, I paid the "pence of ransom" for the privilege of releasing the wheatear from the captivity of the T-trap in which I found it fluttering that summer afternoon on a lower slope of the South Downs.[1]

OF "EARTH-STOPPING" BY SHEPHERDS

By H. Somerset Bullock, 1910

Earth-stopping is the stopping of the fox-earths on nights previous to the day of a meet in the neighbourhood. Shepherds employed in this work generally receive 10s. for every fox which is killed without "getting to ground" in the district under their charge. I recently chatted with a shepherd who told me that he had been an earth-stopper for thirty years, as his father was before him. He remembers, when his father used to return with the usual earth-stopping money, that it was the invariable rule for all of his nine children to have new smocks or new boots. For his part, when he took the job he made a compact with his wife that she should have all the gold he received to save for a rainy day, so long as he retained the odd shillings for spending. They have now a nest-egg of some £50.

OF THE SHEPHERD'S BOTTLE

By The Author

A bottle full of country whig
By the shepherd's side did lig.[2]
Robert Greene,
The Shepherd and his Wife (1590).

The little kegs or wooden barrels in which the shepherds used to carry their "cold thin drink" are still remembered as "bottles." The contents were generally innocent enough. It might be herb beer made from the small dandelion, the burnet, tops of nettles, ginger, sugar,

---

[1] At the chief poulterers in Chichester they know nothing of wheatears—" never heard of them"; at another shop they had a vague recollection of them in the past.—[*Author's Note.*]   [2] Lie.

and yeast; or home-brewed beer of hops, sugar, and ginger; some added a little malt. Others preferred to drink cider. Many of the old poets write of "whig," which was whey or buttermilk. In *The Ancient Drama* it is thus described: "from the whey of milk; after the cheese curd has been separated from the whey by an acid mixture it is called whig, and drunk by the poor classes as beer." The farmers kept these little barrels in various sizes. They were slung on to the "aims" (or hames) of the horses' collars. When a large supply of drink was needed at harvest or shearing time, a shepherd tells me that at such times he has seen as many as a couple of dozen fastened with plaited horsehair to the tail-ladder of a waggon. His wife was horrified to hear that I paid 4s. for my old shepherd's bottle, a quaint little blue (once green) iron-bound barrel, for she had but a few years ago seen three burnt, as out of date and useless. Halliwell thus defines bottle: "a small portable cask used for carrying liquor to the fields (West.). Bag and Bottle, *Robin Hood*, 11. 54."

OF SHEEP-BELLS, ANCIENT AND MODERN

By The Author

The bells still worn to a diminishing extent by cattle and sheep are representative of an extremely early custom. We know from ancient Irish literature that bells for sheep and cattle were employed in Ireland in very early times, and were very similar to what are called "saints hand-bells."[1]

But as the late Sir Henry Dryden pointed out, there is a remarkable difference between the usual form of the early hand-bells and that of bells employed for sheep and cattle. "In the former the mouth part is as wide or wider than the top part or shoulder; but in the latter the mouth is narrower than the shoulder. It may, indeed, generally be assumed that the latter shape is later than the former," though this is of course a different thing from saying that *all* forms of sheep-bell are of more recent date

---

[1] Mr. Anderson names 50 or 60 of these "saints" or "sanctus" hand-bells in Ireland, 6 or 7 in Wales, 2 in England, several in France, and 1 in Switzerland. For a Scottish example see page 271.—[*Author's Note.*]

than hand-bells, a fact which may very well not be the case, and can certainly not be proved. It must be remembered that the earliest cattle-bells were made of wood, and were probably more often of an almost square shape than round, or else rather of the nature of the wooden clappers which may still be seen on the Continent.

About thirty years ago Sir Henry Dryden, in a still

By *Habberton Lulham.*

A "MUSIC-MAKER"

extant letter to the late Sir A. W. Franks, remarked : " The bell which I bought was made for me in Wilts. It is about the largest size made for sheep, and made in exactly the same way as the old bells. I have written a few notes about these riveted bells for Ellacombe's forth-coming book on bells. This is the only way that a smith can make a bell." Sir Henry adds : " This method is still followed for most sheep- and cattle-bells. It appears to have been the usual mode of making hand-bells in the fifteenth, sixteenth, and seventeenth centuries, and perhaps

Shepherds' Arts, Implements, & Crafts

in the eighteenth century, after hand-bells had ceased to be used in the services of the English Church.

"Old bells are of this shape:—

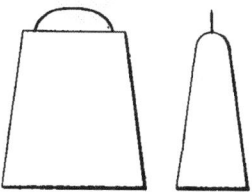

The new ones are:—

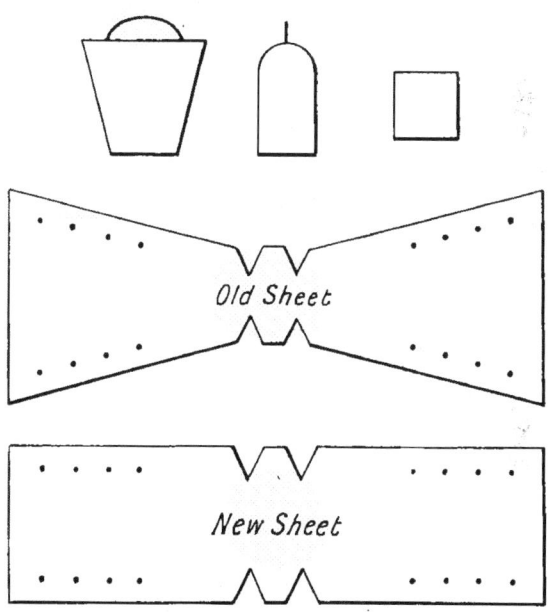

Hence the different form when made up. I believe they were made down to 1650 or 1700 for alarm-bells and other domestic uses. I have six, got about this country, all about 10½ inches high. All these bells were brazed by means of a flux, and were thus made continuous."[1]

With regard to these riveted bells the Rev. H. T.

---

[1] The Rev. J. J. Raven, D.D., F.S.A., in *The Bells of England* (1906), p. 20, has: "This type of riveted bell remained on through many generations, and may be seen now on the Wiltshire Downs as a sheep-bell."

Ellacombe remarks : " My own opinion is that all these little bells, having the appearance of bronze, were formed in the same way that sheep-bells are made to this day. There is a family at Market Lavington, Wilts, by name of Potter, who have made them for generations, and so sonorous are their bells that on a still night they may be heard on Salisbury Plain, at a distance of four miles. Sheet iron is bent into the form, and riveted together. The intended bell is then bound round with narrow strips of thin brass ; some borax is added as a flux, and the whole being enveloped in loam or clay, is submitted to the heat of a furnace, by which the brass is melted and gets intermixed with the heated iron, so rendering it sonorous. Otherwise they were plated with brass, the iron being first dipped in tin, as plated articles of brass are now produced."[1]

In Sussex, sheep-bells were made of smelted or welded charcoal iron of ancient local manufacture. A South Down shepherd puts bells on to a certain number of his flock, and these are allotted their respective places on the Downs, so that he can tell where to find them. One of these shepherds tells me that when driving his sheep he notes the position into which the various members of his flock invariably fall. The same impetuous ones are always to be found far ahead of the rest, on the alert for dainty herbage or any possible mischief. Others keep midway, and some near him. The largest of the welded iron bells, called "clucks" (the most sonorous bells), he puts on to the noted go-a-heads, and if he happens to have any of the small round "musical" ones of bell-metal which cannot be heard at any great distance, he reserves these for the "lazy ones." Some say that it is impossible to live near a farm or sheepwalk where bells are used ; but when they are taken off at shearing-time, and perhaps put aside for a while, the shepherd "misses the music" and wonders who could object to it. In *Rambles in Sussex* Mr. F. G. Brabant remarks : "The tinkling of the bell round the neck of the bell-wethers is a musical sound

---

[1] *Bells of the Church* (Messrs. W. Pollard & Co., by permission).

Shepherds' Arts, Implements, & Crafts

which accompanies the Down rambler along the whole range." The sound of the bells of a flock are especially a help to the shepherd after dark, in a fog, or in case of any attempt at sheep-stealing. A thief fears the noise of the bells of a disturbed flock, as the shepherd will be then on the alert to find out the cause. The old shepherds name the various parts as follows : the wooden neckpiece is the "yoke," the bone or wooden wedge passing through the leather thongs is the "lockyer,"[1] and the small pieces of leather added to make the bell hang level are the "reeders"; the clapper is called the "clipper." The Chichester ironmongers are rarely asked for sheepbells now, but old ones are brought to be repaired. When once a flock-owner has a stock of bells they last for many years, and are used for generations of sheep. A local farmer remarked that they are now too poor to buy them for their sheep.

SOME SPECIMENS OF BELLS.—*See page* 270.

Fig. 1 is a large sheep-bell of Sussex welded iron, brazed or lacquered. Height, 5 inches; circumference at middle 13 inches, at base $12\frac{1}{4}$; side, $4\frac{1}{4}$ across top and 3 at base.

Fig. 2 is a Sussex sheep-bell of welded charcoal iron; a "cluck" said to be of the date of Queen Anne. Height, 4 inches; width at top 5 inches, at base $3\frac{1}{4}$. An old shepherd showed me how, when swung, the sound produced resembles "cluck, cluck"; he knows nothing of the little round bells, having never seen them. The "clucks" can be heard at a great distance on the Downs. On hearing them the shepherds say, "Here come the old cluckers."

Fig. 3 is of the same shape as Fig. 1, but not lacquered, and of a very different tone. I bought it at an old curiosity shop in Chichester, where antiques of metal are a speciality; the owner, who picks up reliable information from his customers and keeps the records in a book, states that it is supposed to be of the reign of George I.

---

[1] The locker, or "lockyer," is sometimes of leather also; see Fig. 6.

Fig. 4 is of bell-metal, cast; inside are the letters R. W. These have been said to stand for Richard Wells of Somerset, a well-known bell-maker of a hundred years ago. But is it not more probable that they should stand for Richard Woodman, who had ironworks at Warbleton in Sussex, in which county it was bought? Mr. T. W. George, of Northampton, remarks in a letter of May 20, 1910, addressed to the writer, that Reading appears to

FIG. 1.   FIG. 2.   FIG. 3.   FIG. 4.   FIG. 5.

SHEEP-BELLS IN POSSESSION OF THE AUTHOR

have been the place of bell-founders who made bells of the globular form, upon which occur the initials G. W. and R. W., and that Richard White was the name of a bell-founder in Reading in 1520.

Fig. 5 is modern, bought at Chichester in 1909, is of bell-metal, cast, "musical." These bells are sold by weight, 1s. per lb., this one weighing ½ lb.

Fig. 6 is from Winchelsea and of the same shape as 1 and 3, showing peculiar fastenings of leather lockyers; height, 4¾ inches; the property of Mr. W. Ruskin Butterfield, the curator and librarian of Hastings Museum and Library.

Shepherds' Arts, Implements, & Crafts 271

Sheep-bells are still made in Birmingham in fairly

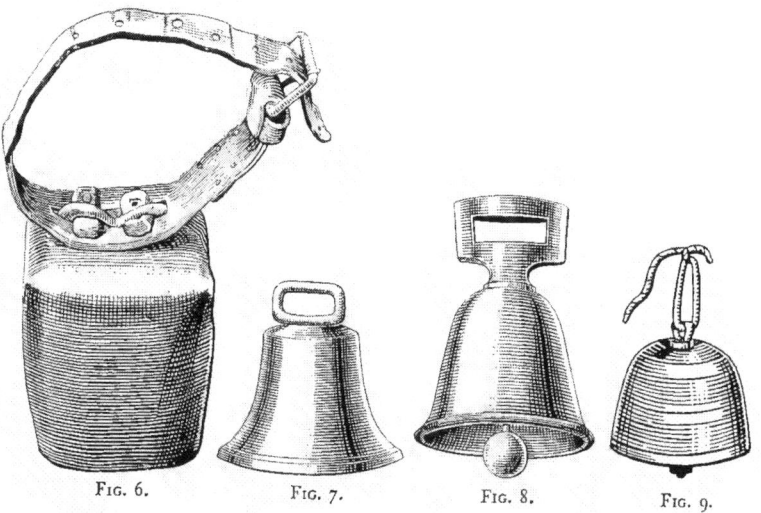

Fig. 6.   Fig. 7.   Fig. 8.   Fig. 9.

large quantities, chiefly for the South American and South

Fig. 10.

African markets (see Figs. 7, 8, and 9). These are mostly

of bell-metal. In their manufacture the loop is cast first ; after casting, the bells are roughly dressed, then barrelled, dipped, and turned on the edge or lip ; lastly, the iron clapper is fixed in position. Sizes vary from $2\frac{1}{2}$ to $4\frac{1}{2}$ inches in diameter.

Fig. 10 represents the "sanctus" hand-bell of Fortingal, West Perthshire, which is here given for the purpose of comparison with sheep-bells. It is of great antiquity, probably dating back to the foundation of St. Cedd's Church (seventh century). Its "tongue" has gone, but it is otherwise in a fair state of preservation.

PLATE I.

THE SIMPLE SUNDIAL OF THE SOUTH-DOWN SHEPHERDS

By EDWARD LOVETT, 1909

An even more interesting survival than the sheep-tally occurs amongst the shepherds of the South Downs. A turf sundial is still to be found in use in a few places from which the cheap watch has not yet driven it. A shepherd, after feeding his flock on roots where they have been "folded" for the night, will take them on to the grassy Downs, returning with them when it is time for the night

Shepherds' Arts, Implements, & Crafts 273

folding. In order to do this he must know at what hour to begin his return journey, for he may have a long distance to go. If without a watch, and with no clocks within hearing, he resorts to one of the turf dials shown in Plate I. If the sun fails him, and his dial consequently does not work, he has to calculate by dead reckoning. In some cases the old shepherds can make very good estimates of the time without either watch, sundial, or visible sun. The form of sundial photographed in Plate I. is made as follows:—Having selected a fairly smooth bit of turf, the shepherd marks a rough circle about eighteen inches

PLATE II

in diameter with a pointed stick, leaving the stick perpendicularly in the ground in the centre. Due south of this he fixes another stick, about twelve inches long, on the periphery of the circle. The south direction is either ascertained at midday by means of another man's watch, or, more frequently, by landmark bearings known to the shepherd. Having done this, he fixes another stick due west, which is, of course, merely a matter of measurement. He then fixes in the intervening quadrant of the circle five sticks for the hours one to five inclusive, so completing a sundial with seven gnomons on its circumference. At three o'clock on an October afternoon, which is about the time shown in the photograph, it may be

T

about time to return to the fold, and the shadow of the third stick from the midday gnomon will then fall on the central stick, and the shepherd will know that it is time to start.

Another form of turf dial is photographed in Plate II. and is much more similar to the ordinary garden sundial. The central stick is the gnomon, and a stick notched for the hours is laid across the ends of the two other sticks pointing due north and due east. I have also seen hour-sticks placed at regular intervals from north to east for the shadow of the central gnomon to fall upon them.[1]

The only reference I have been able to find to the former use of these turf dials by shepherds occurs in Shakespeare's *King Henry VI., Third Part*, Act ii. Sc. v. Perhaps this well-known passage refers to a more elaborate turf copy of the ordinary dial than those above described:

> . . . methinks it were a happy life,
> To be no better than a homely swain;
> To sit upon a hill, as I do now,
> To carve out dials quaintly, point by point,
> Thereby to see the minutes how they run,
> How many make the hour full complete;
> How many hours bring about the day;
> How many days will finish up the year;
> How many years a mortal man may live.
> *Henry VI., Third Part*, Act ii. Sc. v.

AN ANCIENT RUSTIC POCKET-DIAL

By Thomas Quiller Couch, 1862

I have, in my little collection of local curiosities, an old pocket ring-dial, obtained from a labourer in the parish of Pelynt, Cornwall, and a specimen, probably, of

[1] William Aylward of Chichester, who tended sheep on the Downs near this city when a boy, and whose father was a shepherd for many years, invented a turf dial, which for its simplicity beats all others. One now figures on our lawn, and is quite useful. He thus describes his method on the Downs: "On a sunny day with a south wind we could hear the cathedral clock strike, and if the sun was shining we used to fix a short stick upright in the ground and cut a ridge in the turf where the shadow fell, and so on at each hour; and on other days, when the striking of the clock could not be heard, but the sun shining, the dial was ready for use." These shepherds used also to judge the time of day by the position of the sun in respect to the tall cathedral spire.—[*Author's Note.*]

Shepherds' Arts, Implements, & Crafts

an instrument once in ordinary use. Its occurrence is, I believe, rare, but I have met with another, though a defective one, in the possession of a peasant. It is a brass ring, like a miniature dog's collar, and having—in a groove in its circumference—a narrower ring, with a small boss, pierced so as to admit a ray of light. This narrow ring is made movable, to allow for the varying declination

A POCKET-DIAL

of the sun; and accordingly, on either side of it, *i.e.* on the broad ring, is cut in ascending and descending series the initials of the months from June to the December solstice. On the concavity of the great ring, opposite the boss, is engraved a scale of the hours and half-hours. It bears also the inscription:

> Set me right, and use me well,
> And i ye time to you wil tell.

In conformity with this direction, we will, for instance, move the boss on the sliding ring to D (December), and suspend it by the string directly opposite to the sun, when the ray of light, passing through the aperture, will impinge on the concave surface opposite, and tell, with tolerable accuracy, the hour. Shakespeare is the only writer I recollect who alludes to such a form of horologe as having been in common use; and I regard my curiosity the more as I believe it illustrates a well-known passage of our great poet. I am fain to think that it was just such another which gave occasion to the fool in the Forest of Arden to "moral on the time" in words "so deep-contemplative":

> And then he drew a dial from his poke,
> And, looking on it with lack-lustre eye,
> Says very wisely, "It is ten o'clock."

The date of the play of *As You Like It* is generally referred to the year 1600; and as pocket-watches were not introduced into England until about the year 1577, it is very unlikely that the fool would have been possessed of so novel and costly a convenience.[1]

A SHEPHERD'S POCKET-DIAL

By E. B.

The present owner of a curious "timepiece" writes: "A friend presented me with a rude instrument which—as the Maid of Orleans found her sword—he picked 'out of a deal of old iron.' It is a brass circle of about two inches in diameter. On the outer side are engraved letters indicating the days and the months with graduated divisions, and on the inner side the hours of the day. The brass circle itself is held in position by a ring, but there is an inner slide, in which there is a small orifice,

---

[1] Three previous lines should have been quoted as well, thus:

> "I met a fool;
> Who laid him down and bask'd him in the sun,
> And rail'd on Lady Fortune in good terms . . .
> And then he drew a dial from his poke."

This shows us that the sun was out, and the use of a pocket sundial therefore quite possible.—[*Author's Note.*]

Shepherds' Arts, Implements, & Crafts

this slide being moved so that the whole stands opposite the division of the month where the day falls of which we desire to know the time; the circle is held up opposite the sun, the inner circle is, of course, then in the shade, but the sunbeam shining through the little orifice forms a point of light upon the hour marked on the inner side. We have tried this dial and found it gave the hour with great exactness. It seems probable that this was the kind of dial alluded to in Shakespeare's *As You Like It*. 'And then he drew a dial from his poke. . . . "You should ask me what time o' day; there's no clock in the forest," said Orlando.' It was not very likely the fool would have a pocket clock. What, then, was the dial he took from his poke?"[1]

---

[1] Mr. E. Fillingham King, M.A., in his book *Ten Thousand Wonderful Things*, has: "In 1584 watches began to come from Germany, and the watchmaker soon became a trader of importance. . . . Country people, like Touchstone, sometimes carried pocket dials, in the shape of brass rings, with a slide and aperture, to be regulated to the season."—[*Author's Note*.]

SHEPHERDS' PASTIMES

SHEPHERD WITH PIPE AND DOG
(From *A Book of Hours*, 1410)

Tho saugh I stonden hem behinde,
A-fer fro hem, al by hemselve,
Many thousand tymes twelve,
That maden loude menstralcyes
In cornemuse, and shalmyes,
And many other maner pype,
That craftely begunne pype
Bothe in doucet and in rede,
That ben at fèstes with the brede ;
And many floute and lilting-horne,
And pypes made of grene corne,
As han thise litel herde-gromes,
That kepen bestés in the bromes.
  CHAUCER, *The House of Fame*, Book iii.

SHEPHERDS' PASTIMES

In wrestling nimble, and in running swift ;
In shooting steady, and in swimming strong ;
Well made to strike, to leap, to throw, to lift,
And all the sports that shepherds are among.
  SPENSER (1590).

> I can dance the raye, I can both pipe and sing,
> If I were merry ; I can both hurle and sling ;
> I run, I wrestle, I can well throw the bar,
> No shepherd throweth the axeltree so far ;
> If I were merry, I could well leap and spring ;
> I were a man mete to serve a prince or king.
>
> BARCLAY's *Eclogues* (16th century).

A PIPING LAD

By RICHARD JEFFERIES, 1880

A SHEPHERD lad will sit under the trees, and as you pass along the track comes the mellow note of his wooden whistle, from which poor instrument he draws a sweet sound. There is no tune—no recognisable melody ; he plays from his heart and to himself. In a room, doubtless, it would seem harsh and discordant ; but there, the player unseen, his simple notes harmonize with the open plain, the looming hills, the ruddy sunset, as if striving to express the feelings these call forth.

SHEPHERDS' PIPES

By Rev. F. W. GALPIN

In connexion with this subject it is interesting to record the different kinds of pipes which have been used in this country by shepherds and other " herde-gromes that kepen bestés in the bromes," to quote the quaint words of Chaucer.

The true pastoral pipe was a reed-pipe, not necessarily made of reed, but sounded by means of a reed or vibrating tongue. These pipes are of two distinct types, and the earlier of them seems to have been the *double-reed pipe*. To construct this, a small oaten straw was taken when green and one end pressed together with the fingers ; this was then placed between the lips, and the two sides, thus forced together, vibrated on each other under the pressure of the breath. Owing to the slender form of the straw it was sometimes called an oaten "quill," a term also used by the old weavers for a narrow hollow reed.

The double-reed principle, which was known from remote ages in the East, appears in the reed-pipes of the Greeks and Romans. The Western spread of Arabian influence gave a further impetus to its popularity, and the calamus or calamellus became the chalemie, schalmey, or shawm of the Middle Ages. The tube or body of the

By H. Warren.

A LESSON IN PIPING

shawm was made of wood and pierced with seven finger-holes. A similar instrument is still used with the bagpipe by shepherds in Italy and other parts of Southern Europe. (*See illustration*, p. 281.) The shawm has now been replaced by the hautboy, but in both instruments the little pressed straw of primitive times is represented by the double-reed pipe with which they are played. Country children still make these "pypes of greene corne" under the vulgar name of "squeakers," but rarely take the trouble to cut the necessary ventages or finger-holes for a musical scale.[1]

[1] M. Drayton represents Melanthus playing to his sheep, and William Ellis's

The second type is the *single-reed pipe*, a principle which in ancient times obtained a great popularity in Egypt, where it is still seen in the arghool and zummarah of the country-folk. In this case a dry straw or hollow river-reed was selected; at one end, just below the natural knot, a narrow tongue or slip was cut out of the surface of the tube. When this closed end was placed in the mouth, the pressure of the breath caused the tongue to vibrate against the tube and a droning sound was produced. Although it was known in Western Europe throughout mediaeval times, the single-reed principle remained undeveloped except in this country, where as the hornpipe (Chaucer's "cornpype") it was to be found in the pastoral districts.

In Scotland the instrument was called the stockhorn, and in Wales it appeared as the pîb-corn; but in all cases it had much the same form, namely, that of a hollow tube of wood or natural bone pierced with holes for the fingers and with a curved horn attached to the lower end as a bell, whilst into the other end a single reed of straw was inserted and covered by a cap of horn open at the top; this was placed over the mouth and a strong breath set the reed in vibration. Such a form of shepherd's pipe was still in use in Scotland and also in Wales in the last century, and it is supposed to have given the name to the dance called the hornpipe, for which it provided the music.

The full development of the single reed is seen in the modern clarinet, which first appeared in the early part of the eighteenth century. In the bagpipe, which was formerly as popular in rural England as it is now in Scotland, we find a combination of these two principles, the melody pipe or chanter having a double reed, and the drones being furnished with single reeds.

<small>shepherd (1749) "makes his sheep merry and cheers them with songs or else with whistle and pipe," pages 105-106.
  John Dyer, in *The Fleece* (1751), refers to the recent use of pipe and tabor:
  ". . . they bound along, with laughing air,
  To the shrill pipe, and deep remurmuring cords
  Of the ancient harp, or tabor's hollow sound."
[*Author's Note.*]</small>

In addition to the reed-pipes there is another large class called *flute or whistle pipes*, of which the well-known panpipes are both a typical and primitive example. Undoubtedly panpipes were used by pastoral people from the earliest times, but the real whistle-pipe or recorder, as it was afterwards called in England, was not so popular with the shepherd of this country as the reed-pipes, being rather associated, certainly from the days of Chaucer's squire downwards, with persons of quality and estate. The little whistle-pipe with three holes, used with the tabor or small drum, was also heard far more often on the village green and in the May-day revels than on the lonely hills and upland pastures.

Though hardly a pipe, a mention of *the long straight or curved horn* still in use in the mountainous districts on the Continent, and employed in bygone days in this country also, must conclude this cursory note. Formed originally from the branch of a tree, cut in half, hollowed out, and then bound together again with strips of bark, it was blown by the herdsmen to call the cattle. Sometimes from its great length it required a forked prop. The sounds were produced by the vibration of the lips in a cup-shaped mouthpiece similar to that of the trumpet or horn. A shorter instrument, slightly curved and pierced with finger-holes, was popular amongst the herdsmen of the Middle Ages ; in the fourteenth century it was called the cornet, but it has now disappeared in England.[1]

SHEPHERDS AT PLAY

From *The Graphic and Historical Illustrator*, 1834.

"Master," queries Moth, in *Love's Labour's Lost*, "will you win your love with a French brawl ? " On this passage Mr. Douce remarks that the ancient English dance denominated a *brawl* was an importation from France, with which balls were usually opened, the per-

---

[1] Hone, in *The Every-Day Book* (1827), mentions spiral May-horns made of the rind of the sycamore tree as played upon by boys and girls at the weddings of the Southdown shepherds. Some say willow bark was also used for these horns.—[*Author's Note.*]

formers first "uniting hands in a circle"; and then, according to an authority in the "*language François*," printed at Angers in 1579, the leading couple placing themselves in the centre of the ring, "the gentleman saluted all the ladies in turn, and his fair partner each gentleman," the figure continuing until every pair had followed the example set them. . . .

Kiss in the ring yet holds a place among the pastimes of the lower classes in " Merrie England "; [1] and though there is but little probability that the brawl will ever regain its ancient honours in the "Modern Athens," it indisputably once formed the most popular disport of Caledonia, and remnants of the practice are still to be found among the heather. Mr. Douce copies from the *Orchesographie* of Thoinot Arbeau, published in 1588, the music of a Scottish brawl; but we learn from the *Complaynt of Scotland*, printed at St. Andrews forty years previous to the above date, that even at that early period the brawl had become so completely naturalised that it was the ordinary pastoral amusement. The author of the *Complaynt*, speaking of a joyous rural assemblage, says: "They began to dance in ane ring, evyrie ald scheiphird led his vyfe be the hand, and evyrie yong scheiphird led hyr quhome he luffit best." He then proceeds to describe the figure as commencing with "twa bekkis" (nods) and " vith a *kysse*."

SHEPHERDS OF SKYE AND THE REEL OF HOOLICAN

By ALEXANDER SMITH, 1865

(At Mr. M'Ian's Farm)

When Peter came with his violin the kitchen was cleared after nightfall; the forms were taken away, candles stuck into the battered tin sconces, the dogs unceremoniously kicked out, and a somewhat ample ball-

[1] To within the last thirty years kiss in the ring was a favourite pastime on Kew Green, but is no longer allowed.—[*Author's Note.*]

room was the result. Then in came the girls, with black shoes and white stockings, newly washed faces, and nicely smoothed hair; and with them came the shepherds and men-servants, more carefully attired than usual. Peter took his seat near the fire; M'Ian gave the signal by clapping his hands; up went the inspiriting notes of the fiddle, and away went the dancers, man and maid facing each other, the girl's feet twinkling beneath her petticoat, not like two mice, but rather like a dozen, her kilted partner pounding the flag-floor unmercifully; then man and maid changed step, and followed each other through loops and chains; then they faced each other again, the man whooping, the girl's hair coming down with her exertions; then suddenly the fiddle changed time, and with a cry the dancers rushed at each other, each pair getting linked arm in arm, and away the whole floor dashed into the whirlwind of the reel of Hoolican. It was dancing with a will—lyrical, impassioned; the strength of a dozen fiddlers dwelt in Peter's elbow; M'Ian clapped his hands and shouted, and the stranger was forced to mount the dresser to get out of the way of the whirling kilt and tempestuous petticoat.

THE COTSWOLD GAMES

From *The Book of Days*. Edited by ROBERT CHAMBERS, 1869

The range of hills overlooking the fertile and beautiful vale of Evesham is celebrated by Drayton in his curious topographical poem, the *Polyolbion*, as the yearly meeting-place of the country-folks around to exhibit the best-bred cattle and pass a day in jovial festivity. He pictures these rustics dancing hand in hand to the music of the bagpipe and tabor, around a flagstaff erected on the highest hill, the flag inscribed "*Heigh for Cotswold!*"[1] while others feasted on the grass, presided over by the winner of the prize—

---

[1] Cotswold was celebrated for its sheepwalks and "Cotswould lions" (*i.e.* sheep).—[*Author's Note.*]

The shepherd's king,
Whose flock hath chanced that year the earliest lamb to bring.

Drayton's description pleasantly, but yet painfully, reminds us of the halcyon period in the history of England procured by the pacific policy of Elizabeth and James I., and which apparently would have been indefinitely prolonged, with a great progress in wealth and all the arts of peace, but for the collision between Puritanism and the will of an injudicious sovereign, which brought about the Civil War. The rural population were during James's reign at ease and happy, and their exuberant good spirits found vent in festive assemblages, of which this Cotswold meeting was but an example. But the spirit of austerity was abroad, making continual encroachments on the genial feelings of the people; and, rather oddly, it was as a counter-check to that spirit that the Cotswold meeting attained its full character as a festive assemblage.

There lived at that time at Burton-on-the-Heath, in Warwickshire, one Robert Dover, an attorney, who entertained rather strong views of the menacing character of Puritanism. He deemed it a public enemy, and was eager to put it down. Seizing on the idea of the Cotswold meeting, he resolved to enlarge and systematize it into a regular gathering of all ranks of people in the province, with leaping and wrestling as before for the men, and dancing for the maids, and in addition coursing and horse-racing for the upper classes. With a formal permission from King James he made all the proper arrangements, and established the Cotswold games in a style which secured general applause, never failing each year to appear upon the ground himself, well mounted and accoutred, as what would now be called a master of the ceremonies. Things went on thus for the best part of forty years, till (to quote the language of Anthony Wood) "the rascally rebellion was begun by the Presbyterians, which gave a stop to their proceedings, and spoiled all that was generous and ingenious elsewhere. . . ." Drayton is very complimentary to Dover:

Shepherds' Pastimes

> We'll have thy statue in some rock cut out,
> With brave inscriptions garnished about,
> And under written : "Lo! this is the man
> Dover, the first these noble sports began."
> Lads of the hills, and lasses of the vale,
> In many a song and many a merry tale,
> Shall mention thee ; and having leave to play,
> Unto thy name shall make a holiday.
> The Cotswold shepherds, as their flocks they keep,
> To put off lazy drowsiness and sleep,
> Shall sit to tell, and hear the story told,
> That night shall come ere they their flocks can fold.

The sports took place at Whitsuntide, and consisted of horse-racing (for which small honorary prizes were given), hunting, and coursing (the best dog being rewarded with a silver collar) ; dancing by the maidens ; wrestling, leaping, tumbling, cudgel-play, quarter-staff, casting the hammer, etc., by the men. Tents were erected for the gentry, who came in numbers from all quarters, and here refreshments were supplied in abundance ; while tables stood in the open air, or cloths were spread on the ground for the commonalty.

> None ever hungry from these games came home,
> Or e'er make plaint of viands or of room ;
> He all the rank at night so brave dismisses
> With ribands of his favour and with blisses.

Horses and men were abundantly decorated with yellow ribands (Dover's colour), and he was duly honoured by all as king of their sports for a series of years. They ceased during the Cromwellian era, but were revived at the Restoration, and the memory of their founder is still preserved in the name Dover's Hill, applied to an eminence of the Cotswold range about a mile from the village of Campden.

Shakespeare, whose slightest allusion to any subject gives it an undying interest, has immortalised these sports. Justice Shallow, in his enumeration of the four bravest roisterers of his early days, names " Will Squele, a Cotswold man " ; and the mishap of Master Page's fallow greyhound, who was " out-run on Cotsale," occupies some share of the dialogue in the opening scene of the *Merry Wives of Windsor*.

SHEEP-RUNNING ON EXMOOR

By Percy W. D. Izzard, 1910

A shepherd told me how, when a youth, sheep provided him with sport before ever he thought of being associated with flocks for his living. It was when he played the old-time game of sheep-running on Exmoor. It was the custom to take a seven- or eight-year-old wether, shear and grease his tail, and let him loose on one of the hills. Five minutes later about a score of young men would set to catch him, with the object of winning the wether or his value if they could hold him by the tail for one minute. This was anything but an easy matter, for the animal was always kept in and fed up for the occasion, and would run for miles up hill and down dale like a wild stag, with his breathless pursuers behind him. "I went in three times and won each time," said the shepherd; "although once the old sheep ran near eighteen miles with me following for about two and a half hours. His value was reckoned at thirty shillings, and I always took the money in case the sheep died afterwards."[1]

VILLAGE PASTIMES (17TH CENTURY)

From *The Book of Days*. Edited by Robert Chambers, 1869

It is curious to find that shepherds and other villagers in Aubrey's time took part in welcoming any distinguished visitors to their country by rustic music and pastoral singing. We read of the minister of Bishops Cannings, an ingenious man and an excellent musician, making several of his parishioners good musicians, both vocal and instrumental, and they sang psalms in concert with the

---

[1] *Folk-Lore*, 1886, tells us that "In Oxfordshire a fat lamb was chased by girls with tied hands. She who caught the lamb with her teeth was declared 'lady of the lamb.' Next day lamb partly boiled, partly roasted, partly baked, was served to the lady and her companions."—[*Author's Note.*]

organ in the parish church. When King James I. visited Sir Edward Baynton at Bromham, the minister entertained his Majesty at The Bush in Cotefield, with bucolics of his own making and composing, of four parts, which were sung by his parishioners, who wore frocks and whips like carters. Whilst his Majesty was much diverted, the eight bells rang merrily, and the organ was played. The minister afterwards entertained the king with a football match of his own parishioners, who, Aubrey tells us, would in those days have challenged all England for music, football, and ringing. For the above loyal reception King James made the minister of Bishops Cannings one of his chaplains in ordinary. When Anne, Queen of James I., returned from Bath, the worthy minister received her at Shepherd-shard with a pastoral performed by himself and his parishioners in shepherds' weeds. A copy of this song was printed, with an emblematic frontispiece of goats, pipes, sheep-hooks, cornucopiæ, and so forth. The song was set for four voices, and so pleased the Queen that she liberally rewarded the singers.

THE OLD BERKSHIRE REVELS

"In which shepherds took a prominent part."

By L. SALMON, 1909

A century ago, in many parts of England these fairs used to be held with sports and pastimes. They had nothing to do with common "statute feasts," being much more ancient. . . . Perhaps the chief feature and the most exciting of the revels was the backswording. . . . Each village had its champion. The game was played with thick sticks, having a basket upon one end to protect the hand. A wooden stage or platform was erected and enclosed by a rope. The lookers-on stood round, those at the back mounted up in carts and waggons to raise them above those in front. Any one wishing to take part in the game threw his hat into the ring as a challenge. When a head was won there were loud cheers and shouts of "Here's another old gamester." An "old gamester" was one who had won a final prize; a "young gamester" one who had, as yet, broken no heads. A head was counted to be broken when the skin of one of the players had been broken somewhere above the eyebrow and the blood had run down an inch, or when one of the opponents was tired out. . . .

The prize for this "mazing lot of clouting" when all was over was a curious one, more enduring, certainly, than the laurel wreath of the Olympic games, but still of the nature of a crown. It was a tall silver-laced hat. This prize of a hat was, indeed, the prize for all the games at the village revels. They were occasionally given by the squire of the parish, who would sometimes present, as well as the beaver hat, an additional prize of a hogshead of beer; but this extra prize was only an occasional one. The hats were generally considered sufficient reward. They were as much prized and sought after as any Greek laurels. Should there happen to be a man at the revels who seemed to carry an air of importance in his looks and you asked who he was, by your very question you would proclaim yourself unknown to the countryside. There would be pride in the answer, showing that the man did not so value himself for nothing. He and his ancestors had won so many hats that his cottage looked like a haberdasher's shop. What was the origin of this strange prize, and what has become of all these treasured hats? One wonders.[1]

BEDFORDSHIRE SHEARING REVELS

From Hone's *Year Book*, 1832

Anciently at Potton, in Bedfordshire, the wool trade was carried on to a considerable extent.[2] At that period

---

[1] Mr. Blencowe, in 1849, tells us of companies of sheep-shearers in Sussex who were governed by a captain who wore a gold-laced hat, assisted by a lieutenant with a silver-laced one. This may account for some of them. See pages 206, 207.—[*Author's Note.*]

[2] Bedfordshire was famous for its sheep. William Ellis, in his *Sure Guide* (1749), says: "In our part of Hertfordshire we have a notion that the west-country sheep and the Bedfordshire sheep are the two best in England."—[*Author's Note.*]

it was customary to introduce at "sheep-shearing" merry-makings, which were then maintained with a spirit honourable to those engaged in them. A personation of St. Blaise,[1] the reputed patron of the woolcombers, was attended by various characters in gay attire, who performed a rural masque ; and there was a kind of morris dance, with other ceremonies.

> O wassel days ! O customs meet and well !

The "good bishop" was represented by a stripling, dressed in snowy habiliments of wool seated on "a milk-white steed," with a lamb in his lap, the horse, its rider, and the little "lambkin" profusely decorated with flowers and ribbons of all the colours of the rainbow, the latter gaieties being carefully treasured up, and cheerfully presented for the occasion by all who took an interest in its due observance. Imperfect memory cannot supply a minute account of the appearance of the other "worthies" forming this "shearing show" or "revel" as it was termed ; but that their costumes were as diversified and sightly as in the one described above, is as certain as that they were beheld with admiration by the country-folk, for on the festive day

> The neighbouring hamlets hastened there,
> And all the childhood came.

The little town presented an animated appearance for the time being. The "display" has unluckily been long since discontinued. It was, perhaps, the most rural of the many celebrations in honour of the saint once common in manufacturing towns.

---

[1] At Boxgrove, Sussex, "the cathedral of village churches" is dedicated to St. Blaise, and on the modern lamp standards may be seen sharp-pointed crowns formed of woolcombs. In the procession of trades to Kingsland in 1685, "the shearmen and clothworkers had a Bishop Blaise with a mitre of wool, and full-made shirt serving for lawn sleeves."

William Hone, in his *Every-Day Book* (1827), remarks that St. Blaise seems to have neglected the woolcombers. "Since the introduction of machinery by Arkwright and others very little cloth is manufactured by hand. The woolcomber's greasy and oily wooden horse, the hobby of his livelihood, with the long teeth and pair of cards, are rarely seen. When scribblers, carders, billies, and spinning-jennies came into use the wheel no longer turned at the cottage door ; but a revolution among the working-classes gave occasion for soldiers to protect the mills. . . . Time, however, has ended this strife with wool, and begun another with cotton."—[*Author's Note.*]

ST. BLAISE'S DAY IN YORKSHIRE

From *The Book of Days*.   Edited by Robert Chambers, 1869

St. Blaise is generally represented as Bishop of Sebaste in Armenia, and as having suffered martyrdom in the persecution of Licinius in 316. The fact of iron combs having been used in tearing the flesh of the martyr appears the sole reason for his having been adopted by the woolcombers as their patron saint. The large flourishing communities engaged in this business in Bradford and other English towns are accustomed to hold a septennial jubilee on the 3rd of February in honour of Jason of the Golden Fleece and St. Blaise ; and not many years ago this fête was conducted with considerable state and ceremony. First went the masters on horseback, each bearing a white sliver,[1] then the masters' sons on horseback, and their colours ; after which came the apprentices on horseback in their uniforms. Persons representing the king and queen, the royal family, and their guards and attendants, followed. Jason with his golden fleece and proper attendants next appeared, then came Bishop Blaise in full canonicals, followed by shepherds and shepherdesses, woolcombers, dyers, and other appropriate figures, some wearing wool wigs. At the celebration in 1825, before the procession started, it was addressed by Richard Fawcett, Esq., in the following lines suitable to the occasion :

> Hail to the day whose kind auspicious rays
> Deigned first to smile on famous Bishop Blaise !
> To the great author of our combing trade,
> This day 's devoted and due honour 's paid ;
> To him whose fame through Britain's Isle resounds,
> To him whose goodness to the poor abounds,
> Long shall his name in British annals shine,
> And grateful ages offer at his shrine.
> By this our trade are thousands daily fed,
> By it supplied with means to earn their bread.
> In various forms our trade its work imparts,
> In different methods and by different arts ;

---

[1] A lock of combed wool.

Preserves from starving indigents distressed,
As combers, spinners, weavers, and the rest.
We boast no gems or costly garments vain,
Borrowed from India or the coast of Spain ;
Our native soil with wool our trade supplies,
While foreign countries envy us the prize.
No foreign broil our common good annoys,
Our country's product all our art employs ;
Our fleecy flocks abound in every vale,
Our bleating lambs proclaim the joyful tale.
So let not Spain with us attempt to vie,
Nor India's wealth pretend to soar so high ;
Nor Jason pride him in his Colchian spoil,
By hardships gained and enterprising toil,
Since Britons all with ease attain the prize,
And every hill resounds with golden cries.
To celebrate our founder's great renown,
Our shepherd and our shepherdess we crown ;
For England's commerce, and for George's sway,
Each loyal subject give a loud HUZZA.
HUZZA ! [1]

A significant remark is dropped by the local historian of these fine doings, that they were most apt to be entered upon when trade was flourishing.

OLD CUSTOMS AT SHEPHERDS' FESTIVALS

From *The Graphic and Historical Illustrator,* 1834

Great festivals were annually celebrated at the Fountain of Arethusa, in Syracuse, in honour of the goddess Diana, who was fabled to preside over its waters ; and the *Fontinalia* of the Romans were religious observances dedicated to the nymphs of wells and fountains, in which rites the throwing of flowers upon streams and decorating the wells with crowns of flowers formed the chief ceremonies. In our own island this custom has not yet fallen into complete desuetude. Shaw, in his *History of the Province of Morray*, observes that heathenish customs were much practised amongst the people there ; and as an instance he cites that " they performed *pilgrimages to wells,*

[1] *Leeds Mercury,* February 5, 1825.

RETURNING TO THE FOLD.

"Our fleecy flocks abound in every vale,
Our bleating lambs proclaim the joyful tale."—*See page 295.*

and built chapels in honour of their fountains." The practice of *throwing flowers upon the Severn* and other rivers of Wales, alluded to by Milton in his *Comus* and by Dyer in his poem of *The Fleece*, is unquestionably a remnant of this ancient usage. Speaking of the goddess Sabrina, Milton says :

> The shepherds, at their festivals,
> Carol her goodness loud in rustic lays,
> And throw sweet garland wreaths into her stream,
> Of pansies, pinks, and gaudy daffodils.

NINE MEN'S MORRIS AND OTHER GAMES

By The Author

Mrs. Gomme, in her valuable *Traditional Games* (1894), observes that "The following are the accounts of this game given by the commentators on Shakespeare, in that part of Warwickshire where Shakespeare was. The shepherds and other boys dig up the turf with their knives to represent a sort of imperfect chess-board. It consists of a square sometimes only a foot in diameter, sometimes three or four yards ; within this is another square, and so on. . . . One party or player has wooden pegs, the other stones, which they move in such a manner as to take up each other's men, as they are called ; the area of the inner square is called the pound, into which the men are taken and are impounded. . . . These figures are always cut upon the grass, green turf, or leys, or upon the grass at the end of ploughed fields, and in the rainy seasons never fail to be choked with mud." Mrs. Gomme adds that Dr. Hyde thinks the morris, or merrels, was known since the time of the Norman Conquest; of course, the form of the word *proves* Norman origin.

The τριόδιον, with its central "fold" or "mound," and the πεττεία, or board of Palamedes, were probably the originals of merrels and draughts. Ovid, *Tristia*, ii. 477-481, and *Ars Amat.* iii. 157-365, alludes to the men being moved in direct lines, set in a row, and retreating. He says in the former :

> Parva sedet ternis instructa tabella lapillis,
> In qua vicisse est, continuasse suos.

But here there are only three counters on each side. In Ireland it was, and perhaps still is, called "top castle," and played with as many men.[1]

"The ancient game of 'nine men's morris' is yet played by the boys of Dorset. The boys of a cottage near Dorchester had a while ago carved a 'merrel' pound on a block of stone by the house. Some years ago a clergyman of one of the northern counties wrote that in pulling down the wall in his church, built in the thirteenth

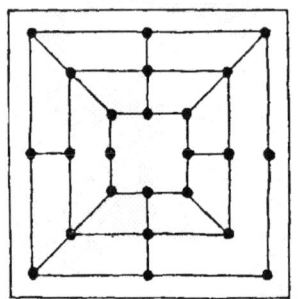

century, the workmen came to a block of stone with 'marrels pound' cut on it."[2]

Joseph Strutt (1836) gives the following account of nine men's morris :—"Nine men's morris is a game of some antiquity. Cotgrave (1632) describes it as a boyish game, and says it was played here commonly with stones, but in France with pawns, or men, made on purpose, and they were termed merelles. It was certainly much used by the shepherds formerly, and continues to be used by them and other rustics to the present hour. But it is very far from being confined to the practice of boys and girls. The form of the merelle-table with the lines upon it, as it appeared in the fourteenth century, is here represented.

[1] See *Notes and Queries*, 1878 (Mackenzie E. C. Walcott).
[2] *Barnes's Glossary* (1864).

Shepherds' Pastimes

"These lines have not been varied. The black spots at every angle and intersection of the lines are the places for the men to be laid upon. The men are different in form or colour for distinction' sake; and from the moving these men backwards or forwards, as though they were dancing a morris, I suppose the pastime received the appellation of nine men's morris; but why it should have been called fivepenny morris I do not know. The manner of playing is briefly this:—Two persons, having each of them nine pieces, or men, lay them down alternately, one by one, upon the spots; and the business of either party is to prevent his antagonist from placing three of his pieces so as to form a row of three without the intervention of an opponent piece. If a row is formed, he that made it is at liberty to take up one of his competitor's pieces from any part he thinks most to his advantage; excepting he has made a row, which must not be touched if he have another place upon the board that is not a component part of that row. When all the pieces are laid down, they are played backwards and forwards, in any direction that the lines run, but can only move from one spot to another at one time. He that takes off all his antagonist's pieces is the conqueror. The rustics, when they have not materials at hand to make a table, cut out the lines in the same form upon the ground, and make a small hole for every dot. They then collect, as above mentioned, stones of different forms or colours for the pieces, and play the game by depositing them in the holes in the same manner that they are set over the dots upon the table. Hence Shakespeare, describing the effects of a wet and stormy season, says:

> The folds stand empty in the drownèd field,
> And crows are fatted with the murrain flock,
> The nine men's morris is filled up with mud.
> *Midsummer Night's Dream.*"

Many of the old people in Chichester know of the game, but say that it is not often played now. I have come across two boards; one is in the stables at Hunter's Race Farm at Lavant, cut on the lid of an old oat-bin.

Farmer Norrel states that it has been cut in his time, and that he has another in his house. It varies slightly from that which Strutt pictures. The squares are three, five, and seven inches respectively; the cross-lines are at the middle of the squares only, and not at the corners. The other is in the cloisters at Chichester, cut on one of the stone window-seats near the south door of the Cathedral. Mrs. Wild, who lived for eighty years in a cottage on the Goodwood estate, remembers seeing the shepherds cut boards in the turf on the downs; and her nephew, who is twenty years younger, tells me that when a boy he used to watch the carters playing in stable-yards while their horses fed. They cut the squares and lines on the stone paving. Seeing Strutt's picture, he at once recognised it as "morrils," and describes the game as "like chess; two could play half the day and neither win." A Sussex shepherd has promised to make me a board; he calls the men sheep, and as they are pounded the name seems suitable enough. A friend, writing from Scotland, says that she remembers playing a similar game there years ago; it was called "dam-brod" (*i.e.* literally "draught-board").

Is it not probable that ninepins was a game invented by shepherds? In Mr. Nelson Annandale's book, *The Faroes and Iceland* (1905), we read about the Farish game of sheep-dogs and of ninepins. "Boys are playing at a game called 'sheep-dogs' on the hill-side. One of them stands above and rolls down a small hoop, made by fitting several rams' horns into one another; another boy stands below, provided with a piece of drift-wood or a small plank borrowed from the carpenter, and strives to hit the hoop uphill; the others chase it when he misses, and bring it back to the bowler. . . . Sheep's horns are used for a variety of purposes . . . they are also set up like ninepins in another game of the same nationality."[1]

Mr. Edward Lovett (an authority on these matters) thinks the above suggestion a good one, adding: "and now might not the ninepin be a symbol of the human

---

[1] By permission of the Delegates of the Clarendon Press.

being?" Certainly Mr. and Mrs. Noah of the child's ark are very like ninepins. Michael Drayton has :

> Then whistle in my fist, my fellow swains to call,
> Down go our hooks and scrips, and we to ninepins fall;
> At dustpoint, or at quoits, else are we at it hard,
> All false and cheating games we shepherds are debar'd.

Dustpoint was a boys' game, in which points or tagged laces were placed in a heap and thrown at with a stone.

THE GAME OF JACK-STRAWS

By William Howitt, 1841

Jack-straws was a great game with us, and if there be any lads that do not happen to know what they are, I will briefly explain them here, because any lad can, at any time, make them for himself. The Jack-straws are a number of straws cut to about three inches long each, or what is better and far more enduring, as many splinters of deal of the same length, and about the thickness of straws, or rather thinner, because they are solid. A lad with his knife may, in a very short time, split off from a thin bit of deal fifty or sixty of these, as well as three or four of twice the length, rounded, and at one end gradually brought to a point, something in the manner of a wooden skewer, only thinner. Three or four or more children may play at Jack-straws, thus :—Let one of the company take up all the Jack-straws in a neat little sheaf in his hand, and holding them about nine or ten inches above the table, let them suddenly fall perpendicularly upon it. They will fall in a tangled heap, and the fun is for each one in turn to remove a Jack-straw from the heap without moving the rest in the slightest degree. Of course, it is easy enough at first, because a few or more of them will be quite apart and disconnected from the heap; but as you proceed the difficulty increases every moment, and a good deal of skill is required to remove some of the Jack-straws, which can only be done by putting the point of the skewer or pointer under one, and lifting it off from the rest by a clever jerk, which, no doubt, gave the original name of jerk-straws, now corrupted to Jack-straws, to the game.

It seems a sport invented by the shepherds to while away the time as they lay on the Downs in summer days. He wins who at the end of the game, which is the entire removal of the heap, has the greatest number of Jack-straws. Any Jack-straws removed by a jerk which shakes some of the others must be thrown back upon the heap again. There are some different modes of playing, but this we have found the most fair and the most agreeable. Some, instead of taking away each of the straws in turn, permit each in his turn to abstract as many as he can without shaking the rest; but this gives a good player who gets the first turn a very decided chance of winning, and often keeps the other players waiting a long time. However, all these laws of the game are subject to the fancy and agreement of those who play, and sometimes one, and sometimes the other, may be tried for variety. A king, a queen, a bishop, and other characters may also be made by dipping the two ends of a Jack-straw into sealing-wax for the king, one end for the queen, and one end into ink for the bishop, which may count four, three, two, or any number agreed upon.[1]

---

[1] These are now sold in the shops, and are called "Spillikins," but are both in themselves, and in the rules for playing, inferior to what are here described.—W. HOWITT.

(Spillikins or spelicans, a name of Dutch origin, from O. Dutch *spelleken*, i.e. a *spillikin*, a small pin or peg.)—[*Author's Note.*]

PASTORAL FOLK-LORE

By Habberton Lulham.

THE SHEPHERD

Alone he bides, a tall old man, and leans
With knotty hands clasping his long ash crook.

HABBERTON LULHAM.

PASTORAL FOLK-LORE

THE SHEPHERD AND HIS LORE

By HABBERTON LULHAM, 1908

ALONE he bides, a tall old man, and leans
With knotty hands clasping his long ash crook ;
His ancient cloak, patched, worn, and weather-stained,
Hangs to his leathern leggings ; at his feet
His two dogs lie, and down the hill below,
In a long sickle line, the feeding sheep
Call in a hundred tones and sound their bells—
Hark to the mellow music ! Sit by him,

And silent though he be from many a year
Of hill-side solitudes, yet as the pine
On yonder crest speaks when the strong wind stirs
Its heart, the breath of sympathy will break
His silentness ; and wiser than he knows,
He hides a world of curious lore behind
Those weather-beaten eyes. Lead him to tell
His tales of dogs and sheep ; of heavy ewes
Frighted by furze-owls or up-springing hares,
And bringing forth strange, beaked, and furry lambs ;
Of how his dogs bark, cowering to the sky ;
And sheep rush panic-stricken when they hear
The witch-hounds in full cry stream overhead,
Hunting some flying soul back to its doom.
And hints has he of arcane mysteries ;
He knows of false dawns, and the hour of flight ;
That cold, dead hour that comes ere night be done,
When dying hearts beat feeblest, and the soul
Most often slips its bars and wings away,
Fanning the air about Earth's sleeping face ;
That is the mystic wind that moves his sheep
To wander a little ; that awakes the larks
To one short flight, and faint, half-hearted song ;
And makes his sleeping dogs uncurl, look forth,
Whimper, and stretch their limbs, and turn and turn
About ere they can rest again ; he tells
How then the upper eastern sky grows light
A space, as if those homing wings broke through
Its leaden grey, or dawn were drawing nigh—
Then, sleep and darkness settle back once more.
And he can tell how down the midnight coombe's
Green, winding hollows, still the little folk
Go dancing 'neath the moon, and round their rings
Sit in applauding circles while their queen,
Light-poised upon a mushroom's milky crest,
Lilts the old fairy laws and spells once more,
Then speeds their quivering wings upon her quests :
And far above them, dark against the sky—
My shepherd tells—late wanderers oft have seen

A ghostly Roman sentinel peer down
From grassy battlements, while in the dene's
Deep, leafy shadows, watching him, the shades
Of British hill-men lurk. But while his tales
Find their slow, plodding words, a smouldering sun
Sinks through the clouds and purple mist behind
The western hills, whereon its last red arc
Glows for a moment like the watchman's fire
Before some ancient camp. He calls his dogs
And sends them forth ; eager they fly to bring
The wandering sheep together ; as he waves
Them on, his crook's head catches the red light,
And shines as when within that Pyecombe forge,
A hundred years ago, his grandsire watched
A cunning hand beat out its long-thought curves.
I will go too, and help him pitch the fold
Down by the hazel-holt, and strew the lines
Of golden swedes. By darkening lanes we wend
Behind the pattering feet and tinkling bells.
It is the hour now of that wondrous blue,
Deep, rich, and luminous, old painters used
To drape about their stately dreams of God ;
That lovely hour between the day and dark,
When all the sky, like some vast jewel, shows
A purple jewel, pure, and ocean-deep,
Set o'er this universe in heaven's floor,
Where through, a little while, the light intense
Filters in soft suffusion to our eyes,
And now the shepherd's lanthorn shines about
His folded flocks, its mellow, orange ray
Making a lovelier, richer blue above
And all around the little ring of light.
Oh, sweet, rare moments fading out so fast !

THE PHYNODDEREE'S[1] SHEPHERDING

By Sophia Morrison, 1909

It was told me by Edward Quayle (aged between sixty and seventy), who lives in the south of the Isle of Man, that "Once upon a time there was a phynodderee living in Colby Glen, who used to do work for a farmer that lived at Ballacrink. Well, one evening it was coming on to snow, and the farmer said to his boys that they had best go and gather the sheep into the fold. The boys went, but found that the phynodderee had got all the sheep in for them, and he had a 'hare'[2] in among them. The boys heard him saying:—'Hiaght mollaght er yn casht veg loaghtan' (My seven curses on the yearling loaghtan); 'she was harder to get in the fold than all the rest; I had to chase her three times round Barrule before I got her driven in.' And when the boys went to the fold in the morning they found the little brown 'hare' lying dead on the ground."

THE LOAGHTAN BEG

By "Cushag," 1908

"Oh, is it a sheep or a witch?" quoth he;
  "Is it only a loaghtan beg?
Or am I awake or asleep," quoth he,
"Or am I the hairy Phynodderee
  That started to catch the meg?"[3]

"I chased her over Barrule," quoth he,
  "And along the side of Clagh Owre;
And three times round Snaefell, like fire, went she
With a screech at the hairy Phynodderee
  That turned the night's milk sour."

---

[1] The popular idea of him is that he is a hairy goblin or spirit. He is said to frequent lonely spots, and is useful to man, or otherwise, as the caprice of the moment leads him.
[2] A little brown Manx sheep.
[3] A hand-reared lamb.

"I have raced the mountain lambs," quoth he,
"And seen them run like deer;
But I never seen wan like yondher," quoth he,
"That could run like the hairy Phynodderee;
She'll not be no right wan, I fear.

"I've seen many sheep in my day," quoth he,
"From the Calf to the Point of Ayre;
But never a wan like that," quoth he,
"Which nearly done the Phynodderee"—
"*Man veg! you have brought me a hare!*"

REMNANTS OF SACRIFICIAL CUSTOMS IN ENGLAND

By WILLIAM HENDERSON, 1879

The Durham butchers mark the sign of the cross on the shoulder of a sheep or lamb after taking off the skin,[1] probably because in the peace-offerings of old it was the priest's portion. In Hunt's *Romances and Drolls of the West of England* (1865) we read: "There can be no doubt that a belief prevailed until a very recent period amongst the small farmers in the districts remote from the towns in Cornwall that a living sacrifice appeased the wrath of God. This sacrifice must be by fire, and I have heard it argued that the Bible gave them warranty for this belief." He cites a well-authenticated instance of such a sacrifice in 1800, and adds: "While correcting these sheets, I am informed of two recent instances of this superstition. One of them was the sacrifice of a calf by a farmer near Portreath for the purpose of removing a disease which had long followed his horses and his cows. The other was the burning of a living lamb to save, as the farmer said, 'his flocks from a spell which had been cast on 'em.'"

---

[1] This is still the custom among some of the trade in Durham; others make a mark like a leaf of bracken. They cannot give any reason for the mark being made. "It is taught them when learning to kill." In some cases in Sussex a butcher who buys from different flocks, but of the same breed of sheep, will mark the carcases differently, so as to be able to distinguish which is which, if necessary, as they may vary in quality.—[*Author's Note.*]

SACRIFICE OF SHEEP AND LAMBS

By the Rev. J. E. Vaux, F.S.A., 1902

The Rev. A. T. Fryer, who was brought up in that county (Devonshire), tells us of a distinctly heathen sacrifice, only modernised, which is still kept up in the parish of King's Teignton, not far from Teignmouth, every Whitsuntide, an account of which is to be found in White's *Devonshire*. It appears that on Whit-Monday a lamb is drawn about the parish in a cart decorated with garlands of lilac, laburnum, and other flowers, and persons are requested to give something towards the expenses of the ceremonial. On Tuesday the lamb is killed and roasted whole in the middle of the village. It is said that it was formerly roasted in the bed of a stream which flows through the village, the water of which had been turned into a new channel temporarily, in order that the bed of the stream might be cleansed. The lamb, when cooked, is sold in slices to the poor at a cheap rate. The precise origin of the custom is forgotten, but a tradition, evidently to be traced back to heathen days, is to this effect :—The village, at some remote period, suffered from a dearth of water, and the inhabitants were advised by their priests to pray to the gods for water, whereupon water sprang up spontaneously in a meadow about a third of a mile above the village, in an estate now called Reydon, amply sufficient to supply the wants of the place, and at present adequate, even in a dry summer, to work their mills. A lamb, it is said, has ever since that time been sacrificed as a votive offering at Whitsuntide in the manner before mentioned. The said water appears like a large pond, from which in rainy weather may be seen jets of water springing some inches above the surface in many parts. The place has been visited by members of different scientific bodies, and whether it is really a spring is still a vexed question. The general opinion appears to be that the real spring is on Haldon Hill, and that after flowing down to Lindridge it loses itself in the fissures of the lime-rock which abounds in the neighbourhood through which it flows; when it meets with some impediment, it bursts up through the soft meadow ground at Reydon, where it has ever had the name of "Fair Water."

Another Devonshire sacrificial custom, evidently having its origin in pagan times, is recorded by an old Holne curate. He says that at Holne, on Dartmoor, the young men, before daybreak on May Day, assemble and seize a ram lamb on the moor. This they fasten to a certain granite pillar, kill it, and roast it whole. At midday they scramble to get slices of it, to secure good luck for the ensuing year. The day ends with dancing, wrestling, and so forth.

SACRIFICIAL CUSTOMS AND OTHER SUPERSTITIONS IN THE ISLE OF MAN

By Sophia Morrison, 1910

Mr. A. W. Moore, in *Folk-Lore of the Isle of Man* (1891), gives an account of an *oural losht* (burnt-offering) in the parish of Jurby in 1880, remarking that even within the last five years there have been several sacrifices, but it is difficult to obtain particulars.

"On May Day eve the people of the Isle of Man have from time immemorial burned all the gorse bushes in the island, conceiving that they thereby burned all the witches and fairies which they believe take refuge there after sunset. The island presented the scene of a universal conflagration, and to a stranger, unacquainted with our customs, it must appear very strange."[1]

It is thus clear that the Manx people placed very great reliance on the influence of fire in protecting them from the power of evil. This influence was also made use of —or would seem to have been made use of—by sacrificing animals as propitiatory offerings to the powers above mentioned. Such a method would naturally be supposed to have belonged to past ages only, if there was not

[1] *Mona's Herald Newspaper*, May 5, 1837.

evidence that lambs have been burnt on May Day eve or May Day *son oural* (for a sacrifice) within living memory. Such sacrifices seem to have been distinct in their purpose from the burnings of animals for discovering witches or driving away diseases, instances of which have also occurred in quite recent times in several parts of England. May 12, *Laa-Bouldyn* (the Beltaine), as it is called in Irish, the first of the great Celtic feasts, was held at the opening of the summer half of the year. Professor Rhys met with some trace of a tradition of sacrifice on this day, an old woman having told him of a live sheep having been burnt in a field in the parish of Andreas *son oural* (as a sacrifice) when she was a "lump of a girl."

On May Day (O.S.) it was a custom to burn a sheep for a sacrifice. Professor Rhys adds in his *Manx Folk-Lore and Superstitions*: "Scotch May Day customs point to a sacrifice having been once usual, and that possibly of human beings, and not of sheep as in the Isle of Man."

An old farmer in the parish of Patrick, Isle of Man, gives me the following charm which has been used by his father at sheep-shearing. It was said when he let go his grip of the sheep :

"Gow magh dy lhome as trooid thie dy mollagh, lesh yn eayn bwoirrin as yn coamrey sonney" ("Go out bare and come home rough, with the she-lamb and the plentiful covering").

Manx people sometimes put into their purses the lucky bone of the sheep. A young woman accidentally dropped one out of her purse before me yesterday. The bone is shaped like Thor's hammer ⚷. I have been told that if a traveller loses his way at cross-roads, not knowing which path to take, he throws the sheep's lucky bone before him, and then follows that path towards which the hammer-end points.

CHARMS AND CURE OF DISEASE BY MEANS OF SHEEP

By THE AUTHOR

Folk-Lore for 1902 has the following interesting passages: "About thirty years ago, when King Edward VII. (then Prince of Wales) was suffering from typhoid fever, it was asserted that the only cure would be to wrap him in a sheep's skin immediately after it had been taken from the animal, while still quite hot, all the wool, of course, being left on. It was believed by many people at the time that this remedy was actually used, and was the means of saving the Prince's life. Only a short time since, during the illness of the Queen of Holland, I heard it referred to as a matter beyond doubt." (*Northampton.*)

"A child had been for some time afflicted with disease of the respiratory organs. The mother was recommended to have it carried through a flock of sheep as they were let out of the fold in the morning. The time was considered to be of importance. The attempted cure of consumption or some other complaints by walking among a flock of sheep is not new. For pulmonary complaints the principle was perhaps the same as that of following a plough, sleeping in a room over a cow-house, breathing the diluted smoke of a limekiln—that is, the inhaling of carbonic acid—all practised about the end of the last century, when the knowledge of the gases was the favourite branch of chemistry." (*Somerset.*)

In the same journal (1908) a Devonshire correspondent gives a cure for whooping-cough: "The child must be taken in early morning to a fold with dew on it, and a sheep be turned off his "form." The child is then rolled in the place where the sheep has been lying." (*Devon.*)

In Northamptonshire and Suffolk a very common charm resorted to for warding off cramp was the patella of a sheep or lamb, known as the "cramp bone," worn as near the skin as possible, and at night laid under the pillow.

Mr. J. Newman of Chichester tells me that the Sussex shepherds used to put the fore-foot of a mole into a little leather bag and wear it round the neck, to keep off cramp; also the galls or excrescences sometimes to be found on beech trees—these are still called "cramp nuts." The shepherds say that only those from beech trees are of any use, but they are scarce, and many an hour may be spent without finding one.

Lady Jane Wilde, in *Legends of Ireland* (1887), reports the following superstition: "When a family has been carried off by fever, the house may be again inhabited with safety if a certain number of sheep are driven in to sleep for three nights."

There may be more sense in this than some think. Wool absorbs infection in a wonderful way. It is to be hoped that the sheep are kept in the wilds for some time after such a venture.

A SHEPHERD BURIAL CUSTOM

By The Author

In *Folk-Lore* (1900) we read that "when a shepherd died it used to be the custom to put a lock of wool into his coffin, the idea being that at the Judgment Day he could thus prove his vocation, which prevented him from being a regular attendant in church. The custom has now become obsolete, but not long ago I heard of a case in which a lock of wool was placed in the coffin of a shearer."

Ann Hickman, of Chichester, aged sixty-eight, whose father was a shepherd at East Ashling, a few miles distant, tells me that she can remember, when he died, seeing a lock of wool put into his coffin. She was a very small child at the time. I fail to find others who even know of the custom. When I asked an old shepherd of Slindon about it he said: "Never heard of it; I don't go to church, can't leave my flock ; but wish I could, it would be a treat. I have four lads under me ; I sends them."

WEATHER WISDOM OF THE SHEEP
By The Author

When the sheep begin to go up the mountains the shepherd says it will be fine weather ; this is always looked for in the Highlands.

The hill-sheep have an instinctive dread of a coming storm.[1] When a moorland shepherd meets his sheep, on a winter's night, coming down from the hill-tops (where they prefer to sleep) he knows that a storm is brewing. Bleating lowly, as if uttering a warning to the younger members of the flock, they seek the shelters on the plains below. An old Scotch rhyme tells us of three lambs that seem to have been slow to obey this instinct, unless perhaps, as they were young, it was not fully developed !

> March said to April :
> " I see three hoggs[2] on yonder hill,
> And if you'll lend me days three,
> I'll find a way to gar them dee."
> The first day it was wind and weet ;
> The second day it was snaw and sleet ;
> The third day it was sic a freeze
> It froze the birds' nebs[3] to the trees.
> When the three days were past and gane,
> The silly puir hoggs cam hirpling[4] hame.

Sir Walter Scott says : " The three last days of March (old style) are called the borrowing days, for as they are remarked to be unusually stormy, it is feigned that March had borrowed them from April to extend the sphere of his rougher sway."

Hone, in his *Every-Day Book* (1826), remarks that " Before storms, kine and also sheep assemble at one corner of the field, and are observed to turn their heads towards the quarter from whence the wind does not blow " ; and the Rev. C. Swainson, in his *Handbook of*

---

[1] See *supra*, " Welsh Sheep," p. 71.
[2] One-year-old lambs. [3] Beaks. [4] Limping.

Weather Folk-Lore,[1] has : "If sheep gambol and fight, or retire to shelter, it presages a change in the weather. Old sheep are said to eat greedily before a storm and sparingly before a thaw ; when they leave the high grounds and bleat much in the evening and during the night, severe weather is expected. In winter, when they feed down the hill, a snowstorm is looked for ; when they feed up the burn, wet weather is near." I saw what Mr. Swainson says, in respect to gambolling and fighting, exemplified at Chichester last winter (1908-9). There had been a long spell of frost, and to a non-expert in weather lore no sign of a change ; all was crisp, blue sky and sunshine. I went into the Palace Meadows, and came upon a delightful scene — the lambs skipping and gambolling, while the sheep with lowered heads fought in a comparatively demure manner. On my return home I told a friend that she must go and see these most fascinating antics. She went the next day ; the frost had suddenly gone, and the sheep and lambs were passive.

Mr. Walter Money, F.S.A., writes : "The old shepherds of Salisbury Plain and of the Berkshire hills, like their forebears, are very weather-wise, and I have heard an old man, who was a perfect mine of local information and folk-lore, render the well-known saying,

> A rainbow in the morning
> Is the shepherd's warning ;
> A rainbow at night
> Is the shepherd's delight,

thus in the local vernacular :

> The rainbow in the marnin
> Gives the shepherd warnin
> To car er's gurt cwoat on er's back ;
> The rainbow at night
> Is the shepherd's delight,
> For thae then no gurt cwoat vill er lack.

A homely way of expressing the famous lines of Byron :

> Be thou the rainbow to the storms of life,
> The evening beam that smiles the clouds away
> And tints to-morrow with prophetic ray."[2]

---

[1] By permission of Messrs. William Blackwood & Sons.
[2] *The Bride of Abydos*, canto ii, stanza 20.

Pastoral Folk-Lore 317

By Habberton Lulham.

FLEECES OF SKY AND LAND

A "lamb's-wool sky." The shepherd looks for rain.

> The unthrift sun shot vital gold
> A thousand pieces;
> And heaven its azure did unfold,
> Chequer'd with snowie fleeces.
>   HENRY VAUGHAN (1621-1695).

I will conclude with the delightful story of the scholar and the shepherd from Hone's *Every-Day Book* (1827). The scholar has not been satisfactorily identified. "One of the *Hundred Merry Tales* teacheth that, ere travellers depart their homes, they should know natural signs; insomuch that they provide right array, or make sure that they be safely housed against tempest. Our Shakespeare read the said book of tales, which is therefore called 'Shakespeare's Jest Book'; and certain it is, that though he were not skilled in learning of the schoolmen, by reason that he did not know their languages, yet was he well skilled in English, and a right wise observer of things; wherein, if we be like diligent, we, also, may attain unto his knowledge. Wherefore learn to take heed against rain, by the tale ensuing of the herdsman that said, 'Ride apace, ye shall have rain.' A certain scholar of Oxford, who had studied the judicials of astronomy, upon a time as he was riding by the way, there came by a herdman, and he asked this herdman how far it was to the next town. 'Sir,' quoth the herdman, 'it is rather past a mile and a half; but, sir,' quoth he, 'ye need ride apace, for ye shall have a shower of rain ere ye come thither.' 'What,' quoth the scholar, 'maketh ye say so? there is no token of rain, for the clouds be both fair and clear.' 'By my troth,' quoth the herdman, 'but ye shall find it so.' The scholar then rode forth, and it chanced ere he had ridden half a mile further, there fell a good shower of rain, that the scholar was well washed and wet to the skin. The scholar then turned him back and rode to the herdman, and desired him to teach him that cunning. 'Nay,' quoth the herdman, 'I will not teach you my cunning for nought.' The scholar proffered him eleven shillings to teach him that cunning. The herdman, after he had received his money, said thus: 'Sir, see you not yonder black ewe with the white face?' 'Yes,' quoth the scholar. 'Surely,' quoth the herdman, 'when she danceth and holdeth up her tail, ye shall have a shower of rain within half an hour after.' By this ye may see that the cunning of herdmen and shepherds, as touching alterations of weathers, is more sure than the judicials of astronomy. Upon this story it seemeth right to conclude, that to stay at home, when rain be foreboded by signs natural, is altogether wise ; for though thy lodging be poor, it were better to be in it, and so keep thy health, than to travel in the wet through a rich country and get rheums thereby."

INDEX

Abingdon, fair at, 237
Agricultural show, at Petworth, 149; of Beds. County Agricultural Society, 150. *See also under* Holkham, and other place-names
Allan, John Hay, *The Last Deer of Beann Doran*, 92
Annandale, Mr. Nelson, *The Faroes and Iceland* (1905), 77-8, 300
Anderson, on "Saints" hand-bells, 265 *note*
Anderson, Col., and old Manx breed of sheep, 65
Anglesey, breed of sheep in, 71
Anglia, breeds of sheep in East, 52-5
Argyllshire, shepherd of, 80
Arkwright, introduction of machinery by, 293 *note*
Aspects of pasture, their marked effect on sheep, 109
Atkinson, Rev. J. C., *Forty Years in a Moorland Parish* (1892), on the hamp, 245
Antiquary's Portfolio, on ewe-milking, 95; on dress of fifteenth-century labourer, 246
Antiquity of the sheep, Prof. Rolleston on, 69
Aubrey, John, on the South Downs, 37-40
on village pastimes, 290-91
on wearing of straw hats, 248
"Auld Kep," a Border collie of the old type (twice winner of the International Cup), 154
Author, articles by the, 27, 36-7, 53-5, 63-6, 72-3, 76-8, 95-6, 105-7, 109-11, 112-14, 118-20, 127-30, 143-4, 167-9, 173-5, 181-5, 193-4, 210-11, 219-21, 235-6, 245-51, 257-60, 264-72, 297-300, 313-19

"Backstays" worn by shepherds of Romney Marsh, 52
Bagpipes, played by English shepherds, 217, 258, 283, 284
Baynton, Sir Edward, visited by James I. at Bromham, 291
Bedfordshire, famous for sheep, 292 *note*
shearing revels, 292-3
Bell music on the South Downs, 10
Bells, old and new shapes of, 265-7. *See also* Sheep-bells, Saints bells, etc.
Bell-wether's fleece, a shepherd's perquisite, 214
Berkshire revels, the old, 291 *seq.*
sheep and shepherds in, 46
Birmingham, class for sheep-dogs at (1860), 126; sheep-bells still made at, 271
Blackface breed of sheep, 75, 91-2, 101
called "collies," 130
Blackmore, Stephen, a famous shepherd, 20 *note*
Bogie or *buggie*, a sheep's skin, 100; buggie-flaying (=taking off a sheep's skin whole), 100
Bolg, a sack or bag, 100
"Bone-eater," sheep called the, 118
"Border Leicesters" (sheep) found in Ireland, 75
Bottle, the shepherd's, 264; specimen in author's possession, 265
Branding of lambs in Skye, 185-7
Bridget, St., anniversary of, formerly called Ewe-milking Day, 77
Britons, plaiting preceded weaving among ancient, 226
"sheep-counting scores" may be derived from ancient, 194-200

Shepherds of Britain

Bronze age, sheep in the, 69
Brookside (shearing) Company, 206
Broom-squires, as sheep-shearers, 206
Buck-horn, used by Scottish shepherds, 93
Buggie-flaying. *See* Bogie
Builing, in the Shetlands and Orkneys, 97, 100
Burel, a coarse native fabric in ancient England, 226
Burial custom, shepherd's, 314
Burnt-sacrifice, survival of in modern England, 309
"Bush, shepherd's," how to form a, 18-19
Bustards (called *wild turkeys*) on South Downs, 17 ; on Salisbury Plain, 38

Carnarvonshire, sheep character in, 71
Carving of crooks by shepherds, 258-9
Cast, *i.e.* overturned, a term used of sheep, 59 *note*
Celtic numeral system, 199
Charles II., clipping-time customs under, 213
Charm used at sheep-shearing (Isle of Man), 312
Check or chequered patterns used by ancient Britons, 226
Cheviot breed of sheep in Shetland and Orkney, 101
Chichester, crooks for shepherds made at, 258
nine men's morris or merelles at, 299
weather wisdom of sheep exemplified at, 316
wheatear trade obsolete at, 264 *note*
Chipping Campden, indebted to wool trade, 230
Church, shepherds' dogs in, 167 *seq.*
See also Dog-noper and Dog-whipper
Circencester, indebted to wool trade, 230
Clifford, Henry, "The Shepherd Lord," 56-8
Clipper, the clapper of a sheep-bell (Sussex), 269
Clipping, 68 (*note to illustration*), 211-12, 213, 218. *See also under* Coke

Clipping-time, *i.e.* sheep-shearing time, 212-19 ; "to come in clipping-time," 212
Cloth trade, in different parts of England, 225-36. *See also under place-names*, Kendal, Lincoln, Leominster, Norwich, Worcester, etc.
in Ireland, 227 ; in Flanders, 228.
See also "Wool trade"
Cluck, a large iron sheep-bell (Sussex), 268-9
Clynnog Fawr (Wales), dog-tongs at, 173 *note*
Coke, Thomas William (Earl of Leicester, 1837), inaugurates "Coke's Clippings," 214; worked among his own shepherds, 215
"Collie," disputed meaning of, 129
Communal ownership of sheep, 53
"Corn-pipes," used by Scottish shepherds, 93
Cornish "hair," duty remitted by Black Prince on, 234
Cornwall, breed of sheep in, 28-9
burnt-sacrifice survivals in, 309
snails believed to be eaten by sheep in, 117
Cotswold breed of sheep, 4, 71, 287, *seq.*
games, the, 287 *seq.*
lions, *i.e.* sheep, 287 *note*
"Cottons," Kendal, 232-3
Counting-out games, elementary methods of reckoning used by children, 200
"Cramp-bone" (*i.e. patella*) of sheep, used for curing cramp, 313-14
"Cramp-nuts" (excrescences on beech trees), belief of Sussex shepherds as to, 314
Crook, shepherd's, 9, 16 ; form in various counties of, 255-8, etc.
handles, carved by shepherds, 258
made out of an old gun barrel, 255, 257
Pyecombe, 9, 257, 258
Crues (Shetland), 189
Cumberland sheep-farming, 60

Dancing by shepherds, 94, 106, 286-7
Dartmoor, May Day custom of roasting ram on, 311
Deer, driven out by sheep, 92

Index

Derbyshire, sheep of, 55
Devonshire, sheep and shepherds, 28 ; sheep-cure for whooping-cough practised in, 313. *See also* Dartmoor, etc.
Dial, ancient rustic pocket, 274-6. *See also* Sundial
"Dishley," or "New Leicester" sheep, 56 *note*
Dog-noper, formerly a church official, 174
Dogs, used for hunting down sheep, 98 ; behaviour of in church, 169
Darwin on powers of South American sheep-dog, 138
French sheep-dogs to have iron nail-studded collars for fighting wolves, 105
of white colour thought best for sheep, 105
classified as sheep-dogs and collies, 127 ; (*a*) English and Sussex sheep-dogs, 22, 27, 128, 129, 143, 144, 147, 165 ; how trained, 147 ; (*b*) collie-dogs or collies, 13, 16, 79, 127, 130-31, 132-4, 136 *seq.*, 140, 142 ; how trained, 148-9 ; trials of, 149-67 ; behaviour in church of, 167 ; James Gardner's sayings on, 88-9
Dog-tongs, or "lazy tongs," used for expelling dogs from church, 170-173
Dog-whipping, church official appointed for, 173-5
Dorchester, nine men's morris, or merrels, still played by boys at, 297
Dorsetshire sheep and shepherds, 27 ; shepherds' crooks in, 258. *See also under* Dorchester
Drayton, Michael, on dress worn by shepherds (sixteenth century), 247
on leading of sheep by shepherds, 106
on "Lemster ore," 236
on playing of a shepherd to his sheep, 283
on shepherds' games, 300
Dress, black, in ancient Ireland, 67, 238 ; in ancient Scotland, 235, 238 ; in ancient England, 226, 245 ; at various periods in England, 245-52 ; of shepherds, 238-9, 241, 245, 248. *See also under* Hat, Hood, Pouch, etc.
Dress—
smock-frock, how worn, 248-50 ; colours in various parts, 251
of Sussex (blue, drab, and grey), 16
of Wilts (blue), 40
of Herts (olive green), 251
of Northamptonshire, 108
of Shropshire, 249
Drought, terrible effect on sheep of, 114-16
Dudeney, John, a learned shepherd, 19-20, 21 *note* ; engaged in wheatear trade, 263
Durham, butchers' mark on sheep-carcases at, 309
Dustpoint, game of, 300

Eagles and shepherds, 17, 83
Ear-marking in England (seventeenth century), 182
Earth-stopping by shepherds, 264
East aspect of pasture, effect on sheep, 109
Edward III., encourages wool-trade, 4-5
Edward VII. (when Prince of Wales), sheepskin cure believed to have been used in illness of (Northampton), 313
Ellis, Alexander, on the "sheep-counting score," 194, 200
English or Sussex sheep-dog, his points described, 127 ; introduced into Isle of Man, 143
Ennerdale, Herdwick sheep of, 60
Environment, affects shepherds, 6
Epitaph on Highland shepherd, 80
Eskdale, ear-marks at, 183
numerals of, 199
Ewe-milking, 77, 95-6. *See also* Milk, etc.
Ewe-milking, ancient Irish name for the 1st of February, 95
Ewes, fate of overturned, 17
weep for their lambs, 111
Exmoor breed of sheep, 28
"Sheep-running" on, 290
Extremes (of climate or food) fatal to sheep, 109

Fairs of sheep and cattle. *See under* place-names

Faroe Isles, sheep of, 66; *rueing* practised in, 66 and 220
Findon, fair at, 126
Flemish weavers introduced into England, 241
Flock, as part of the landlord's property, 61
Flotwhey. *See under* Milk (Scottish methods of preparing)
Flounders, a disease fatal to sheep, 113
Fortingal, West Perthshire, "sanctus" hand-bell of, 272
Foster-mothers for lambs, 112
Foxes and sheep, 17, 72, 81, 83
Fox-hunting welcomed by shepherds, 83
France, account of shepherding in, 105-6; sheep led by shepherd in, 105-6; game of merelles in, 298-9
Fustean scones, eaten by Scottish shepherds, 94

Games, the Cotswold, 287 *seq.*
Garb, shepherds'. *See* Dress, shepherds'
Gardner, James, of North Cobbinshaw, Midlothian, shepherd and collie-dog trainer (born 1840), 84-9, 137
Garlanding of sheep and rams, 217
Gatesgarth Fells, no longer "stinted pasture," 61
Glasgow, Irish flock-masters ship wool to, 76
Grainger, or *Granger*, explained, 213 *note*
Grasmere, ear-marking at, 182
Greek double-reed pipes, 283
Grey cloths of Kent, 234
Gyffylliog (Wales), "lazy-tongs" used for expelling dogs at, 170, 173

"Hair," Cornish, 234
Halifax, wool trade established at, under Henry VII., 5
Hamp, a kind of smock-frock, 245-6
Hampshire, sheep and shepherds of, 26
Down sheep in Ireland, 76
white smock worn in, 251
Hand-bells, "saints" or "sanctus." *See* "Sanctus"
Hare, a little brown Manx sheep, apparently mistaken for a "hare" by the "*Phynodderee*" (Isle of Man belief), 308
Haslock. *See Hawselocks*
Hat, tall silver-laced, worn by captain of shearers, 292
Hawselocks, i.e. neck-locks, the wool about a sheep's throat, 212
Helvellyn, range, habits of sheep on, 61
Henry VII., wool trade under, 5
Herding of sheep, almost unknown in Shetland. *See under* Shetland
Herdwick breed of sheep, 60-62; wool of Herdwicks bred in Wales turns white, 61
Heredity, its effect on shepherds, 8-10
Herrick, on "Lemster ore" (cloth), 236
Hertfordshire, Bedfordshire sheep in. *See* Bedfordshire
Hibbert, Dr. S., on public sheep-marking in Shetland, 188; on wool of Shetland sheep, 221
Highlands, weather omens from sheep in, 315
sheep and shepherds of. *See under* Scotland
"Hirsel," a flock or company of sheep, 135
Hogg, a one-year-old lamb, 315
Hogg, James (the "Ettrick Shepherd" or "Mountain Bard"), 28, 140-42, 176-8
Holkham shearing feast, 214, 217
Holne, sacrifice of ram lamb on May Day at, 311
Home, attachment of sheep to their, 72
Hone, William, *Every-Day Book*, on Mayhorns, 285
(1826), on weather wisdom of sheep, 315
(1827), on weather wisdom of shepherds, 318-19
(1827), on St. Blaise and the woolcombers, 293 *note*
The Table Book (1827), 92-3, 218
Year-Book, 292-3; on labourers' dress, 248
"Honeycomb" pattern, on smocks, 250
Hoods, worn by Scottish shepherds, 93
Hoolican, the reel of, 286-7
Horn, the long, 285
Hornpipe, the, a single-reed pipe, 284

Index

Houl'ers, i.e. "holders," applied to dogs that "held" sheep for their masters, 144
Hunt, *Romances and Drolls of West of England*, on survivals of burnt-sacrifice in Cornwall, 309
Hunting of sheep with dogs, 98
Huts of shepherds on South Downs, 17

Iceland, breeds of sheep in, 66-8, 78
   rueing practised in, 66, 220
Ill, the leaping, a disease of sheep, 113
Ilsley, East, importance of sheep fair at, 46-9
Instinct in sheep, 70, 71, 83, 100
Inverness-shire, shepherd-bard of, 80
Ireland, cloth manufactures in, 76, 241; Irish cloth used in England, 227
   dress in ancient, 67, 238
   men of, called "Westmen" in old Norse history, 77
   "ploughing by the tail," practised in, 220
   rueing practised in, 220
   sheep-farming in, described, 75; various breeds of sheep in, 75; ancient form of sheep-house in, 76; importance of sheep-farming in ancient, 76-8; sheep-bells in ancient, 265; connexion between ancient Irish and Faroes breed, 67, 77-8; shepherds of, 73-4, sheep-dogs of, 73, 142
Isle of Man, breeds of sheep in, 62-6; sheep-marking in, 184-5; sheep-bells not used in, 78. *See also* Loaghtan
   burnt-sacrifice in, 311
   superstitions in, 308

Jack-straws, game of, 301
Jakobsen, Dr., *Shetland Norn Dictionary* by, 190
   on Shetland ear-marks, 190
Jaloff method of reckoning numerals compared with Welsh method, 198
Jumpers (sheep), how dealt with, 120
June, the sheep-shearing month, 206
June, Rosebuds in, a well-known shearing song, 207-8, 210
Jurby (Isle of Man), burnt-offering at, 311

Keeir sheep, 64
Keep, out to, 53
Kendal cloth, sold at Stourbridge fair, 237
   "cottons" (so-called), 232-3
   "Green" (cloth), 233, 235
Kent, grey cloths of, 234
   sheep and shepherds in, 49-52
Keswick, "clipping" customs at, 218-19
"King" of shepherds, 288
   and queen of shearers, 217
King's Teignton (Devon), survival of burnt-sacrifice at, 310
Kingussie, shepherd's plaid made at, 238 note
   Shetland sheep at, 131
Kirk, shepherds' dogs in, 167 *seq*.
Kirn milk. *See* Milk (Scottish methods of preparing)
Kiss in the ring, 286
Kitts, sour. *See* Milk (Scottish methods of preparing)

Lady of the lamb (Oxfordshire), 290
Lake Country shearing customs, 219
   sheep-farming in, 60. *See also* Ear-marking, etc.
Lamb, burnt alive in Cornwall, 309
   sacrificed in Devon at Whitsuntide, 310
Lambs, branded in Skye, 185-7
   chased by girls in Oxfordshire, 290
Land's End, sheep-feeding near, 29
"Laughton" breed of sheep. *See* "Loaghtan" (*also* "Lughdoan")
Lavant, Shepherd Stacey of, 113
Lead-mines, said to affect sheep-pasture, 114
Leaping ill, the, a disease of sheep, 113
Led by the shepherd, sheep, 105-6
Leeds, wool trade established in Henry VII.'s reign at, 5
Leicester breed of sheep, 55, 56 *note*, 65, 71, 109; in Shetland and Orkney, 101
Leominster (or "Lemster") cloth, commonly called "Lemster ore," 236
Lilting, or milking tunes in ancient Ireland, 77
"Lincoln Green" (cloth), 236
Lincolnshire "longwools," 55; in Ireland, 75
   drovers, 107

326 Shepherds of Britain

Lincolnshire, sheep-counting score in, 194
Liver-fluke, a disease fatal to sheep, 113
"Loaghtan," or "laughton" breed of sheep, 62, 63; colour and points of, 63-6
Loaghtan Beg, poem of the, 308
Lock of wool placed in shepherd's coffin, 314
Lockyer, part of a sheep-bell's fastenings (Sussex), 269
London, white smock worn in counties round, 251
Looker, i.e. watcher or shepherd (Romney Marsh), 51
"Lucky bone" of the sheep, superstitions regarding, 312
Lug of bonnet, spoon carried by Scottish shepherds in, 94
Lughdoan, correct spelling of "laughton," 64

Manx breed of sheep. See Isle of Man sheep-dog a "holding," not a driving, dog, 143. See also under Dog
Mark, private, of farmer stamped on sheep with pitch, 219. See also under Ear-marking, Branding, etc.
Marsh-pennywort (Isle of Man) poisonous to sheep, 113
Matterdale range, habits of sheep on, 61
May Day or May Day eve, lambs or sheep burnt in Isle of Man on, 312; ram sacrificed at Holne on, 311; probable ancient sacrifices of human beings in Scotland on, 312
Mayo husbandry, former cruelties of, 220
Meeting, agricultural, at Holkham, 214-15; on Woburn estate, 215
Meeting-time of shepherds, 59
Meg, a hand-reared lamb (Isle of Man), 308
Merelles. See Morris, nine men's
Merino breed of Spanish sheep, 71
Merle, a variety of the collie-dog, 127
Midlands, ancient breed of sheep in, 69
blue smock worn in, 251
modern breeds of sheep in the, 55

Migratory habits of sheep in Carnarvonshire, 71
Milk, Scottish methods of preparing, 94. See also Ewe-milk
Milking tunes in ancient Ireland, 77
Morrils. See Morris, nine men's
Morris, fivepenny, 299
nine men's, or merrels, 297-300
Mothering required of shepherds, 147
Mount up, Sussex sheep-dog trained to, 129

Native sheep. See under place-names, e.g. Shetland, Orkney, Isle of Man, St. Kilda, Hebrides, Faroes, Ireland, etc.
Neolithic times, question as to presence of sheep in England in, 69
Nick, uses of the word, nicks of tally-sticks, 193
Ninepins, game of, 300
Nope = to knock on the head, e.g. dogs (North). See Noper
Noper, 174-5
Norfolk breed of sheep, 53-5
North aspect of pasture, effect on sheep, 109
Northamptonshire, beliefs as to sheepskin cures in, 313
"Cramp-bone" of sheep used for cures in, 313
a drover of, 108
smock-frock worn in, 108
Northumberland, breed of sheep in, 97
Norwich: a centre of cloth industry, 227
Notches, of tally-sticks, 193
used for "scoring" at cricket, 193
Nottinghamshire, breed of sheep in, 55

Old settlers, when sheep are called, 70
"Ore." See Leominster or "Lemster"
Ore-weed, or oar-weed (properly woar-weed), sheep fed on, 29
Orkney, breeds of sheep in, 101. See also under Shetland
sheep-mark in parish register of, 189
Orphan lambs, foster-mothers found for, 112
Oural, a sacrifice (Isle of Man), 311, 312
Out to keep, meaning of, 53

Index

Ovis cauda brevi, or short-tailed sheep, 97
Oxfordshire, "lady of the lamb" in, 290

Paab, a small yard built of uncemented stones, 64
Pack system of wages, 74
Panpipes used by shepherds in early times, 285
Pasture, effect on sheep of, importance of aspect of, 109-10
  effect of lead-mines on, 114
  poisonous, 112-13
  profitableness of, in sixteenth century, 4
  stinted, or common land, 61
Patella of sheep used for curing cramp, 313-14
Pennygrass or pennywort, poisonous to sheep, 113
Pentads, Welsh method of reckoning by, 198
Petworth agricultural show, 149
Phynodderee, the (Isle of Man), 308
Pibgorn (Wales). *See* Hornpipe
Pindar, or pinder, the keeper of a pinfold, or pound for cattle, 97 *note*
Pipes, shepherd's, 93, 282-5
"Piping lad," a, 282
"Plaid, shepherd's," 238
Plaiting, in prehistoric times preceded weaving, 226
Playing of shepherd incites sheep to feed, 106
Ploughing *by the tail* in Mayo, 220; near Cavan, *ibid. note*
Pocket-dial, an ancient rustic. *See* Dial
Poisonous sheep-pasture, 112
  to sheep, plants, 113
Portreath, calf sacrificed by farmer at, 309
Portland breed of sheep, 27
Portslade Shearing Company, 206
Potton, Bedfordshire, ancient shearing-revels at, 292
Pouch, shepherd's, 79
Presentiment in sheep, 70-71
Punding, in the Shetlands, 64, 97, 98, 189
Pyecombe crook. *See* Crook

Quirk, preserver of old Manx breed of sheep, 65

Rabbits, killed and eaten by sheep, 118
Rachael, weeping of the Fleecy, 111
Ram, the black, meaning of, 209
  lamb sacrificed upon Dartmoor at May Day, 311
  white, a shearer's feast, 208
Rams, garlanding of, 217
"Rasp," a collie-dog belonging to Mr. James Gardner, 87, 137
Raths, or circular stockades of the ancient Irish farmer, 77
Ravens and sheep in Carnarvonshire, 72
Recognition by sheep after shearing, difficulty of mutual, 112
Reeders, accessories of a sheep-bell, 269
Revels, shepherds', 285-96
Reydon, sacrifice of lamb at Whitsuntide at, 310-11
Reyme. *See* Milk (Scottish methods of preparing)
Rigwelted, meaning of, 59
Roman double-reed pipes, 283
  graves in England, shears found in, 219
Romans, originally a race of shepherds, 38
Romney Marsh, sheep and shepherds, or "lookers," of, 51-2
Rosebuds in June, a famous Sussex song, 207-8, 210
"Rows" at Stourbridge fair, 237 *note*
Rue, to pull the loose wool off sheep, 219-21
Rueing: in the Faroe Isles, 66, 220; in Iceland, 66, 220; in Mayo, 220; in Orkney and Shetland, 66, 98, 99, 189, 220-21
Russia, sheep of modern, compared with native Shetland sheep, 97, 98
Ryeland breed of sheep, 71

Sacrificial customs in England, survival of, 309
St. Blaise celebrated in Yorkshire, 294
St. Columb, sheep of, 28
St. John's, ear-marks at, 183
St. John's Vale, "clipping" customs at, 218
St. Kevern, sheep of, 28
St. Kilda, breed of sheep in, 66, 68

Shepherds of Britain

Saints bells and sheep-bells compared, 265-72
Salisbury Plain, sheep and shepherds of, 34 *seq.*, 316
Saltersbrook (Yorks), shepherds' meeting at, 59
" Sanctus " hand-bells, 265, 272
Saunders, David, shepherd of West Lavington, 37
Sayings on dogs (by James Gardner), 88-9
Scandinavia, sheep of modern, 97, 98
" Scat," or land-tax, once paid to Denmark in cloth by Shetlanders
Scatholds, wild sheep of the, 97
Scent of sheep as a means of mutual recognition, 112
Scholar, of Oxford and weather-wise shepherd, the, 318-19
Scilly Isles, sheep of, 29
" Score " used in counting sheep, 24, 192-200
Scotland, sheep, shepherds, and dogs of, 73, 79-101 ; in Hebrides, 66, 68, 78. *See also under* Shetland and Orkney
sheep-bells not used in, 78
Sea-weed, sheep feed on, 97, 99
Shap, Herdwick sheep of, 60
Shearers' king and queen, election of, 217
Shearing customs in Sussex, 206, 207-11, 292 *note* ; revels in Bedfordshire : *see* Potton
mutual recognition by sheep after, 112
Shears found in Roman graves in England, 219
Sheep, ancient English breeds of, 67 ; black breed of the Faroes and Iceland, Ireland, Shetland, and Hebrides, 78 ; four-horned, 78 ; remains of, hardly distinguishable from those of the goat, 69 ; short-tailed breed of Northern Europe, 66, 97, 98 ; three-horned, 78
as a moderate drinker, 119
courage in, 28
deer driven out by, 92-3
exported from England to Spain, 3-4
formed part dowry of Catherine Plantagenet on marriage to Henry III. of Spain, 3

Sheep, garlanding of, reason for the custom, 217
great numbers of, in charge of drovers, 107
hunted down with dogs trained for the purpose, 98
kill and eat rabbits, 118
led by the shepherd. *See* Leading of sheep
immense numbers in England of, from the thirteenth century onwards, 3-4
profitableness of, in the sixteenth century, 4
quality of, affected by soil, 109-10
terrified by a dog's bark, 146
weather wisdom of, 315-19
Sheep-bells and " saints bells " compared, 265-72 ; in Ireland, 78 ; absent from the Isle of Man, 78 ; not used in Scotland, 78
-carcases, marks made by butchers on, 309
-counting " score," 194-200
-fairs. *See* Fairs
-farming, encouraged by Edward III., 4-5 ; stimulated by wool trade, 3
-house, ancient Irish, 76
-*killing*, a name given to pennywort, 113
-lore (superstitions about sheep), 305-19
-marking, 181-90
-pasture, ploughed up for crops, 43
-pool, 147. *See also* Sheep-washing
-running on Exmoor, 290
-shearing, 204-21 ; charms used at, 312
-skin cloak worn by shepherds in ancient Britain, 245
-skins, tanned and made into waterproof cloths in Shetland, 241
-stealing, 176-8
tun, 52
-washing, 147, 203-6 ; decline of, 211-12
-*wash*, a festival in the North, 213
Sheffield, small demand for crooks at, 257
Shepherd—
Aylward, Charles, 26
Blackmore, Stephen, 20

Index 329

Shepherd—
  Clifford, Henry (the "Shepherd Lord"), 56-8
  Dalgleish, Wattie, 131
  Dudeney, John, 19-21, 263
  Gardner, James, 84-9, 137
  Garlow, David (tale of the trial course), 160-67
  Hogg, James ("Ettrick Shepherd"), 176-8
  Mackenzie, Alastair, 80
  Piper, George, 26
  Saunders, David, 37
  Smith, 118
  Stacey, 113
Shepherd-boy to land-holder, 24
Shepherd burial custom, 314
  communal or town shepherd at Lydd, 53
  king, election of, 217, 288
  lore, 305
  "Shepherd Lord." *See* Clifford, Henry
Shepherdesses, 20, 21, 106, 107
Shepherds, English. *See under* Shepherd; Irish, 73-4; Scottish, 79-80; Welsh, 70
  born to the craft, 8-9
  contentment of, 11-12
  earth-stopping by. *See* Earth-stopping by shepherds
  effect of environment on, 6
  effect of heredity on, 8-10
Shepherd's bottle. *See* Bottle
  bush, description of a, 18
  plaid, 238-9
Shepherds' customs, traced by Aubrey in many cases to Roman origins, 39
  feasts, revels, sports, and games, 281-302
  huts, 17
  meeting-time, 59-60
  wages, 11-13, 39, 59; peculiar form of on Gatesgarth Fells, 61; "pack" system of, 74
Sheppey, meaning of the name, 50
Shepsters, women employed as sheep-shearers in Scotland and anciently in England, 212, 260
Shetland, breeds of sheep in, 64, 66, 78, 131
  long reddish hairs on a rare sheep in, 221
  wild sheep of, 97, 98

Shetland, herding and housing o sheep almost unknown in, 97 98; *builing* and *punding* in, *ibid.*
  rueing in, 66, 220
  shawl-wool, 66
  "tweeds," 240
  wool, extraordinary fineness of, 239
  wool, Hibbert on, 221
Short-tailed sheep, 97, 98
Shots, lambs of inferior qualities, 186
Shropshire breed of sheep, 65
  breed of sheep in Ireland, 75
  smock worn in, 249
Sieves made from perforated sheepskins, stretched on hoops, 100
Singing of shepherd incites sheep to feed, 106
"Sirrah," one of Hogg's famous sheepdogs, 140-41
Skeat, Professor, letter on the "sheep-counting scores," 195
Skiddaw, Herdwick sheep of, 60
Skin of sheep, used as cloak by fishermen, 100
Skye, lamb-branding in, 185-7
  shepherds of, and the "reel of *Hoolican*," 286-7
Sligo County, poisonous pasture, 113
Slings employed by shepherds, 259-61
Sliver, a lock of combed wool (Yorks), 294
Smell the wind, sheep able to, 71
Smock, or smock-frock, wearing of the, 16, 40, 108, 248-51. *See also under* Dress, shepherds' various colours of, 251
Snails, eaten by sheep, 29, 116-18
Snowdonian range, sheep of, 72
Soil, quality of, affects sheep, 109
Somerset, an idle shepherd of, 29-34
  illness believed to be cured by walking among sheep in, 313
Songs sung by wives of Scottish shepherds, 94
South aspect of pasture produces fine-woolled sheep, 109
Southampton, wool-ships in the sixteenth century sail from port of, 4
Southdown, sheep and shepherds, 3-26, 55, 242
  sheep-fair, 14-17
  sheep-walks, 13
South Downs, shepherds' huts on, 17
  sundial made by shepherds on, 272

Southey, Robert, *Commonplace Book*, quoted on ear-marking, 182-4
Shepherds' Guide quoted by, 182
Merino breed of sheep, 71
"Spear-half" and "spindle-half," 235
Spillikins, game of, 302
Spindle, emblematic of the female sex, 235
Spoon carried by Scottish shepherds in *lug* of bonnet, 94
"Squeakers," 283
Staggers, the, produced by poisonous pasture, 113
Stinted pasture, 61
Stockhorn (Scotland). *See* Hornpipe
Stone implements, collected by Stephen Blackmore, 20 *note*
Storms fatal to shepherds, 83
Stourbridge, the great fair of, 237-8
Strayed cattle, announcements as to, formerly made in church, 182
sheep, returned to owners at shepherds' meeting-time, 59
Suffolk breed of sheep, 53-5
"cramp-bone" of sheep used for cure of cramp in, 313
dog, contests International cup at Perth, 160, 163, 165
Sundial, 272-7 ; simple form of, used by South Down shepherds, 272
Superstitions about sheep, 305-19
Surrey, "sheep-counting score" in, 194
Sussex, Pyecombe crook of, 257
shearing customs in, 292 *note*
sheep and shepherds of. *See under* South Downs
sheep-bells from, 268-70
"sheep-counting score" in, 194
sheep-dog beats a Scottish rival, 129 ; points of, 127
sheep-washing in, 205

Tail, ploughing by the, 220
Tally, signification of, 191 *note*
Worcestershire, for lambs, 191
-stick registers, 191
Tar-marks on sheep, 185
Tenterden, weaving practised at, 234
Thacker, Old Bill, of Gedney, 107
Thor's hammer, sheep's "lucky-bone" shaped like, 312
Threlkeld, ear-marks at, 183
Throckmorton, Robert, wins the famous coat-making wager, 241, 242

Toll-bars, formerly a great obstacle to droving, 107
Tongs, lazy. *See* Dog-tongs
"Top-castle," Irish game of, 297
Towen, a sand hillock (Cornwall), 29
Training sheep-dog pups, method of, 147
Trematode, the liver-fluke (a parasitic worm), 117
Trials, sheep-dog, at Tring, 149 ; at Caithness, 149 ; at Petworth, 149 ; in Bedfordshire, 151 ; in Wales, 151 ; on the Highland Border, 152 ; at Perth, 154, 155 *seq.*, 160-67
"Turk," one of James Gardner's dogs, 87
Turkeys, sheep-dog pup trained with, 148
wild, *i.e.* bustards, on South Downs, 17

Umbrellas, immense, once used by shepherds, 40
"Unwearable" Irish tweeds, *i.e.* cloth that cannot be worn out, 76

Veg (Manx) = vagrant or vagabond, 309

Wages, shepherd's, in Yorkshire, 59
"pack" system of, 74
Wakefield, wool trade established in Henry VII.'s reign at, 5
Wales, care of sheep in ancient, 69
habits of sheep in, 69
Walton, near Wetherby, clipping-time at, 213
Warwickshire, sheep in, 55
shepherds' crooks in, 258
Wash-pool, for sheep, 203-6 ; rights as to, 204
Water, strong repugnance of sheep against entering, 204
Weather wisdom of sheep, 315-19
Weddings, objection to dogs at, 172
Weights for holding corn made from sheep-skin, 100
Welsh shepherds, and pasturage, 70, 109
West aspect of pasture produces fine-woolled sheep, 109
West Dean, shepherds' crooks made at, 258
Weyhill, great fair at, 146

Index

"Whaff," 163, 169
"Wharry," one of Mr. James Gardner's collie-dogs, 138
Wheatears, formerly caught in immense numbers on South Downs, 20
shepherds' traps for, 261-4; trade in, 261-4
Whig, i.e. whey or buttermilk, 264-5
Whipping of dogs out of church. *See* Dog-whipping
White colour recommended for sheep-dogs in France, 105
Whitsuntide, lamb sacrificed in Devon on, 310
Wild, Mrs., on shepherd's waterproof calico cloaks, 251
Wiltshire breed of sheep, 34-45
sheep fairs, 43
Wiltshire, white smock worn in, 251
Winchester, ancient fair at, 237
Winter's Tale, sheep-shearing feast described in, 213
Wishart, Edward, sheep-marking warrant granted to, 189
Woburn shearing-feast, 215, 217
Wolves, precautions against, recommended to shepherds in France, 105-6
Wonersh (Surrey), blue cloth made at, 236

Wool, abundance of, in England in fifteenth century, 228
lock of, placed in shepherd's coffin, 314
Woolcombers, 293, 294
Wool-packs, at Stourbridge fair, 238
"Wool-scraps," a perquisite for gatherers, 204
Wool trade, in England in fourteenth to eighteenth centuries, 229-36
sheep-farming stimulated by, 3-4
staple manufacture of England, 4
under Henry VII. and Elizabeth, 5
with France, 229; with Netherlands, 4
Worcester cloth, excellence of, 227
lamb tally, 191
"Worm" in sheep's foot, supposed, 100
Worsted, in Norfolk, centre of a new method of making woollen cloth 227; famous for its manufactures, 227
Wreck, Spanish, sheep supposed to be introduced through, 62

Yoke, neck-piece of a sheep-bell (Sussex), 269
Yolk of the fleece, used in washin sheep, 212
Yorkshire, sheep and shepherds of, 56-60

THE END

Made in the USA
Middletown, DE
27 August 2016